LIFE MOVEMENTS IN PLANTS

LIFE MOVEMENTS IN PLANTS

Sir. Jagadis Chunder Bose

LIFE MOVEMENTS IN PLANTS

Sir. Jagadis Chunder Bose

ISBN : 8888006782593

First Published 1931, reprint 2017 in India by

Facsimile Publisher
12 Pragati Market
Ashok Vihar, Ph-2
Delhi-110052, India
E-mail: books@facsimilepublisher.com

Transactions of the Bose Research Institute, Calcutta, Vol. VI.
1930–1931

LIFE MOVEMENTS IN PLANTS

EDITED BY

SIR JAGADIS CHUNDER BOSE

M.A., D.Sc., LL.D., F.R.S., C.S.I., C.I.E.

CORRESPONDING MEMBER, ACADEMY OF SCIENCES, VIENNA; MEMBER OF INTERNATIONAL COMMITTEE OF INTELLECTUAL CO-OPERATION, LEAGUE OF NATIONS; FOUNDER AND DIRECTOR, BOSE RESEARCH INSTITUTE CALCUTTA

WITH 80 ILLUSTRATIONS

LONGMANS, GREEN AND CO.

LONDON • NEW YORK • TORONTO

1931

CONTENTS

LIFE MOVEMENTS IN PLANTS

I.—INTRODUCTORY

BY

SIR J. C. BOSE

THE plant-world affords a unique opportunity for studying the stages by which a simple and primitive organ becomes gradually transformed into one of greater complexity. The evolutionary process has been active not only in morphological differentiation, but also in the development of specialised mechanisms for the performance of the various vital functions, though they have not attained the same degree of the differentiation in the plant as they have in the animal. It should be borne in mind that the nomenclature of the active tissues is determined by their function and not by their structure. 'Muscle' is the name given to the specialised contractile tissue, while 'nerve' designates the tissue for conduction of excitation to a distance. However, there still exists the long prevalent conception that the physiological mechanisms of animal and plant are fundamentally different because they have been evolved along divergent lines. The results of the experimental investigations that have been carried out in my laboratory during the last thirty years suffice to prove that the mechanisms as well as the life-processes in the plant and in the animal are essentially similar.

Two of the main conclusions resulting from my long-continued investigations in the physiology of plants relate to the ascent of sap and to the transmission of impulse. Certain critics have urged that both these phenomena can be accounted for as being the outcome of purely physical factors. I have, on the contrary, adduced evidence that they are the expression of the activity of the living protoplasm, inasmuch as they are arrested by conditions which would not affect the play of physical forces.

Since the publication of the detailed results given in my previous works on these subjects, many new observations have been accumulated at my Institute by my scholars and staff which are described in the pages of this number of the Transactions. The issue, in a single sentence, is between the physiological and the physical. It is difficult to see what further evidence can be demanded than that here presented in support of the physiological standpoint. The ascent of sap and the transmission of impulse have been shown to be dependent for their activity upon external conditions which have long been known to be favourable to the contractility and irritability of animal tissues, and the activity of these processes is accompanied by movements and electrical variations which correspond with those accompanying the activity of muscle and nerve. It seems impossible that there can be any other logical conclusion from the facts than that these physiological activities of the plant are essentially similar to those of the animal, all of them being manifestations of the fundamental properties of the living protoplasm of which both plants and animals consist. If this conclusion be rejected, there is nothing left but the other horn of the dilemma, that the recognised views on the physiology of animal muscle and nerve, based on results obtained by similar experimental methods, are no more to be accepted than are those which are here urged with regard to plants.

In correlation with the sessile habit of the plant, the reactions of its contractile and irritable tissues are in general less vigorous than are those of the muscle and nerve of the locomotor animal. In order to observe them it has been necessary to devise recording apparatus of high precision and sensitiveness and automatic in action. All error arising from personal bias has thus been completely eliminated by the plant being made itself to record the inner workings of its life.

Several such instruments have been employed in the researches detailed in these papers. But many of the experiments described, notably those dealing with the ascent of sap by the method of the erectile response of the leaf, are so elementary that the main facts can be observed with the simplest apparatus or, indeed, without any at all. And yet these elementary experiments yield results which can only be reasonably interpreted as evidence of fundamental physiological activity.

The following accounts of the recent investigations carried out by myself and my scholars are arranged in the following order, and I take the opportunity to prefix a short note indicating the principal aim and scope of each of these groups of researches.

The Ascent of Sap

(*a*) Characteristics of the erectile response of leaves.
(*b*) Methods of record of the erectile response of the leaf.
(*c*) Effect of poison on the ascent of sap.
(*d*) Effect of the temperature of the water of irrigation on the rate of ascent of sap.

The experimental results indicate that the transport of sap is essentially due to the vital pulsatory activity of a propulsive tissue.

The next series of papers relates to further investigations on the phenomena of

Rhythmic Pulsation in Plants

(*a*) The relation between the pulsatory movement of Desmodium and the ascent of sap.
(*b*) Discovery of rhythmic pulsatory movement in other plants.

The next series of papers deals with

THE TRANSMISSION OF IMPULSE IN PLANTS AND ANIMALS

(*a*) The transmission of impulse in both ascending and descending directions.
(*b*) Excitation of plants by polar action of electric current.
(*c*) Polar excitation of animal nerve under electric current of increasing intensity.
(*d*) Effect of physiological block in arresting the conduction of impulse.
(*e*) The partial and complete block of conduction of impulse by chloroform.
(*f*) Effect of change of environment in modifying the internal physiological activity of the plant.

The solution of the highly complex problem of the tropic effect of light on organs which exhibit pulsatory movement has been attempted in the paper on the effect of light on oscillating organs.

The extension of the work of the Institute in the direction of bio-chemistry is due to Professor N. C. Nag, the Assistant Director, who, assisted by the scholars in the department of chemistry, has contributed the following papers :

(*a*) The Proteolytic Enzymes of *Carica Papaya* (I).
(*b*) The Proteolytic Enzyme in the Seeds of *Phaseolus Mungo*.
(*c*) The Proteolytic Enzymes of *Carica Papaya* (II).
(*d*) Chemical Examination of some Indian Medicinal Plants.

II.—THE ASCENT OF SAP

BY

SIR J. C. BOSE

THERE has long prevailed a consensus of opinion that the rise of sap in the stem and leaves is a purely physical process. The various investigations which I have carried out on the subject during the last quarter of a century indicate, on the other hand, that it is essentially a physiological process effected by the rhythmic activity of living cells throughout the length of the plant, in and from the absorbing root to the transpiring leaf. It is the vital activity of the living cells that causes the propulsion of sap, since it takes place even in the absence of root-pressure and of transpiration.

It should be pointed out that for a satisfactory understanding of the phenomenon it is necessary to determine the character of the cellular activity underlying the ascent, how that activity is initiated, and by what means a definitely directioned transport of sap is maintained. I have already shown elsewhere that it is not the mere presence of living cells, but their pulsatory activity which causes the propulsion.[1] In regard to the directive movement of sap it has been further shown that this is determined by a peristaltic action similar to that in the animal. The peristaltic wave in the animal is found to follow the stimulation gradient, that is to say, from the point activated by stimulation to the unstimulated region. Similarly the direction of the propulsion of sap is determined by differential activity at the two ends of the organ, the flow proceeding

[1] *Plant Response* (1906); *Comparative Electro-physiology* (1907); *Physiology of the Ascent of Sap* (1923).

from the more to the less activated region, whether upwards or downwards.[1]

New Method of Investigation

The problem of the movement of the sap is one of the most debated in plant-physiology; the obscurity of the subject is, in a large measure, due to the presence of numerous co-operating factors, and it has been difficult to disentangle the essential from the non-essential elements in the problem. There are certain crucial tests, however, by which the physiological can be discriminated from the physical factors.

In the present investigation a new method has been perfected, namely, that of the erectile response of the isolated leaf, the speed of which gives an approximate measure of the rate of rise of sap. This method has, in fact, been rendered more sensitive than in the experiments in which I have previously employed it. The experiments are direct and simple, and the very tangible results obtained offer most convincing proof of the physiological nature of the phenomenon.

Before proceeding further it is desirable to make a clear statement of the physical theory and its implications, as contrasted with the observed physiological reactions.

Physical Theory

According to the generally held physical theory, the motive power for movement of sap is attributable to root-pressure, supplemented by backwardly transmitted suction caused by the transpiration from the leaves as expressed in the following quotation: 'It is certain that the water is not merely driven upwards from the root or the base of the stem, by root-pressure acting like a force pump, but that the removal of water from the conducting channels exerts a force transmitted backwards as far as the absorbing organs, causing in these a corresponding entry of water.'[2]

[1] *Motor Mechanism of Plants* (1928), pp. 292–346.
[2] Pfeffer, *Plant Physiology*, English translation, p. 208.

As regards the effectiveness of osmotic forces in the rapid transport of sap, Strasburger came to the conclusion that 'Osmotic forces act too slowly to be of any value, and, moreover, there is no fixed distribution of osmotic substances that would account for such a current.'[1] The very marked effect of heat and cold on the rate of ascent of sap, which will be described, also affords conclusive evidence against the theory of osmotic action as the essential factor.

A rise of sap may occur in the absence of root-pressure, as in the case of isolated stems and leaves when water is supplied to their cut ends. In such cases the rise of sap is supposed to be entirely due to the backward suction produced by evaporation at the transpiring surface of the leaf. If this physical theory be correct, then

(1) The temperature of the water supplied at the cut end of the leaf should have no effect on the rate of rise of sap.

(2) The rate of movement of sap should be unaffected by chemical substances which would modify physiological activity.

(3) Abolition of transpiration should be followed by arrest of the rise of sap, since the motive power for transport is supposed to be solely due to transpiration by the leaves.

Physiological Reactions

As a matter of fact, however, all the above-mentioned conditions do actually affect the ascent of sap, which is evidence that it is a physiological and not a physical process. The following are the characteristic variations which accompany change of conditions:

1. *Variation of internal pressure.*—A certain amount of internal pressure is necessary for the initiation and maintenance of rhythmic activity. This is seen in the renewal of pulsation in the quiescent heart of the Snail, when the intra-cardiac pressure is increased. Similarly,

[1] Strasburger, *Text-book of Botany*, English translation, p. 187.

when water is withheld from *Desmodium gyrans* the leaflets of the plant cease to pulsate, the activity being renewed on irrigation (*cf.* Experiment 43, p. 78). So again, when a plant is subjected to drought, its internal hydrostatic tension and turgor become diminished with the resulting depression or arrest of cellular propulsive activity as shown in the decreased rate of the ascent of sap.

2. *Variation of temperature.*—The pulsatory activity of the animal heart is increased, within limits, by a rise of temperature. The rhythmic activity of a Desmodium leaflet is also found to be quickened under a thermal rise; thus, while the period of a single pulsation was 4·75 minutes at 23° C., it was 1·25 minutes at 38° C. The rate of the ascent of sap has also been found to be similarly modified when the water for irrigation is raised from a low to a moderately high temperature and *vice versa.*

3. *Chemical agents*, such as depressants or poisons, cause corresponding modification in the rate of ascent of sap.

4. *Electric stimulation.*—The pulsatory activity of a Desmodium leaflet is arrested under strong stimulation and is revived after a period of rest. If the ascent of sap be essentially due to rhythmic activity of living cells, then the ascent should be retarded or arrested under electric stimulation, the ascent being renewed after an interval of rest.

5. *Abolition of transpiration.*—If the movement of sap persists even after the abolition of transpiration, it would prove the existence of an independent cellular activity for the propulsion of sap.

The above are some of the crucial tests by which the physiological reaction of living cells can be definitely discriminated from mere physical action. These tests will be employed in the investigations described in this and in the following chapters.

Physiological continuity in the stem and leaf.—The propulsive activity is manifested in the stem, in the petiole, and in the leaf along its midrib. The leaf will, however, be chiefly employed not only for its special advantages for the investigation, but also for the establishment of a wider generalisation.

The general principle of the method of erectile response is to take an organ, be it a stem or a leaf, and subject it to drought, which causes a movement of drooping. On irrigation there is a responsive erectile movement of the various drooping organs ; the rate of the erectile movement gives an indication of the activity of the ascent of sap. This is exhibited in the following striking experiment.

The Revival of a Dying Plant

Experiment 1. *Revival of a dying plant.*—An isolated leaf of the Cornflower was subjected to considerable drought,

Fig. 1. Photographs of three dried-up and twisted leaves of Cornflower (*Centaurea lanata*) (left series).
Full revival and erection after application of warm water for 2 minutes at the cut end (right series).

in consequence of which it exhibited collapse ; the leaf was dried up, twisted and crumpled, being apparently dead. Application of warm water (which enhances pulsatory activity in general) to the freshly cut end of the leaf now produced the most remarkable transformation. As water was absorbed and pumped up, the dying leaf exhibited its revival first by untwisting, and then by rearing itself up erect in a time as short as 2 minutes. The revival of the leaf when shown by shadow-projection is a very striking sight. The photographic reproductions given in fig. 1 exhibit the remarkable transformation.

Experiment 2. *Revival of a dying stem.* The experiment is best carried out with herbaceous stems which possess very marked activity of propulsion of sap. Among these may be mentioned *Centaurea lanata*, *Centaurea moschata*, and Antirrhinum. An isolated cut stem of *C. lanata*, which is here briefly designated as Cornflower, soon droops, and its leaves collapse, falling one over the other. Application of warm water to the freshly cut end of the stem causes lively movements on the drooping leaves (fig. 2).

The physiological explanation.—Physical theory cannot offer any satisfactory explanation of this and the following experiments. The physiological explanation is, however, simple and satisfactory. According to it, the cells which have been rendered inactive on account of insufficient hydrostatic distension, become activated at the irrigated end by the absorption of water. It has already been stated that the law which determines the direction of propagation is that the sap flows from the end that has been activated to the inactive end. In the present case the lower end was activated by supply of water, and the resulting flow was from the lower to the upper end. However, by reversing the turgor-tension gradient, it is possible to reverse the direction of the flow of sap (*cf.* Experiment 7).

Rôle of transpiration.—When the shoot and the root of the plant are equally turgid, the activities at the two ends balance each other, with the result that there is no marked directive movement of sap. The transpiring leaves do not raise water by exerting a sucking force as such, but their transpiration assures differential activity in the plant by causing incipient drought in the transpiring region of the shoot. The turgor-tension gradient extends from the root to the apex of the stem; it is the differential physiological activity induced at the two ends that drives the sap upwards against the force of gravity. There is

FIG. 2. Drooped stem and leaves of Cornflower under drought (left figure).

Revived and erected after application of warm water, H, at the cut end (right figure).

a limit to the height to which water can be raised by physical action; but there is no such limit in the case of physiological propulsion from cell to cell.

Effect of Poisoning

Experiment 3. *Collapse of erect stem under poisoning.*—In sharp contrast with the revival of a dying leaf by the stimulating action of warm water is the collapse of an

Fig. 3. The fully erect stem and leaves of Cornflower (left figure).
Poisonous solution P causes a complete collapse (right figure).

erect and vigorous shoot of Cornflower when a poisonous solution is applied at the lower cut end. The effect of this is that the leaves, from below upwards, fall down in succession; the stem also undergoes collapse and the plant soon becomes a huddled mass of dying tissue (fig. 3).

It should be remembered in this connection that a poisonous solution can travel through a killed tissue, owing to the suctional activity of the living cells situated higher up in the organ.[1] The activity underlying the propulsion of sap, as already stated, is not confined to any particular region, but exists throughout the length of the plant. In the experiment with poison, successive zones of the stem

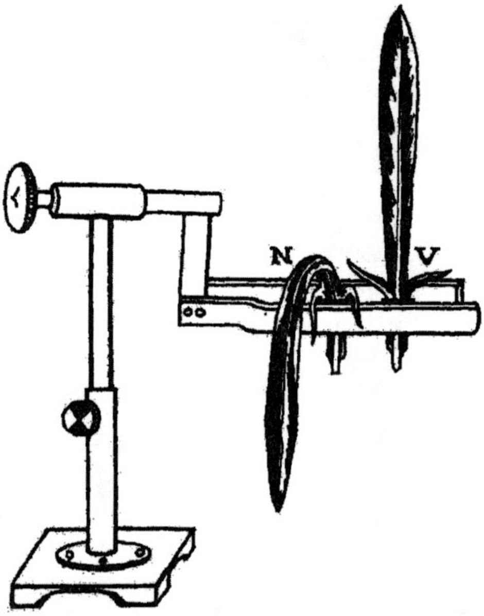

FIG. 4. Abolition of transpiration by vaseline.
Uncoated leaf N, originally erect, fell vertically down within a few minutes; the vaselined leaf V remained erect for more than two days.

were put out of action; this resulted in the final abolition of ascent, as will be demonstrated later (*cf.* Experiment 29).

It is not, however, necessary to take so drastic a step as poisoning to prove the activity of living cells in the propulsion of the sap; for there are other methods by which the existence of a vital factor is demonstrated, as described in other papers in this volume.

The rise of sap, as already stated, has been supposed to

[1] *Physiology of the Ascent of Sap* (1923), pp. 22, 78.

be mainly due to the backward suction caused by the transpiring leaves. This would appear to be highly improbable from the results described in Experiment 1. In that case the leaf was so dry that transpiration from its surface was practically impossible. In spite of this the leaf after irrigation became revived and erected itself in a short time. The next series of experiments was undertaken in order to find out whether the movement of sap persisted even after the complete abolition of transpiration, which was effected by coating the surface of the leaf with an impermeable varnish. I first tested whether freshly boiled vaseline proved to be an efficient coating for the purpose.

Experiment 4. *Vaseline as an impermeable varnish.*—Two fresh erect leaves were held side by side on a supporting clip; the one to the right, V, was vaselined, while the other, N, was left uncoated to serve as a control. The uncoated leaf at once began to droop and doubled over vertically downwards in the course of half an hour or so. The varnished leaf, on the other hand, remained perfectly erect for more than 48 hours (fig. 4), proving the effectiveness of the coating in preventing loss of water by transpiration.

The Independent Propulsive Activity

Experiment 5. *Rapid erection of vaselined leaf after irrigation.*—A wilted leaf of Cornflower was coated with vaseline, and suitably mounted in a bottle. On application of warm water to the freshly cut end of the midrib, the leaf which was hanging vertically downwards became fully erected in the course of about 3 minutes in spite of the fact that the added weight of vaseline opposed the erectile movement. The experiment proves the existence of a propulsive activity which is not dependent on transpiration.

The following experiment further demonstrates the existence of this activity in the living leaf.

Experiment 6. *Difference of response in a poisoned and in an ordinary leaf.*—Two similar leaves of Cornflower were taken, one of which was killed by half an hour's immersion in a solution of formaldehyde. The killed and the living leaves were made to droop to the same extent by exposure to the sun, after which the killed leaf, A, was

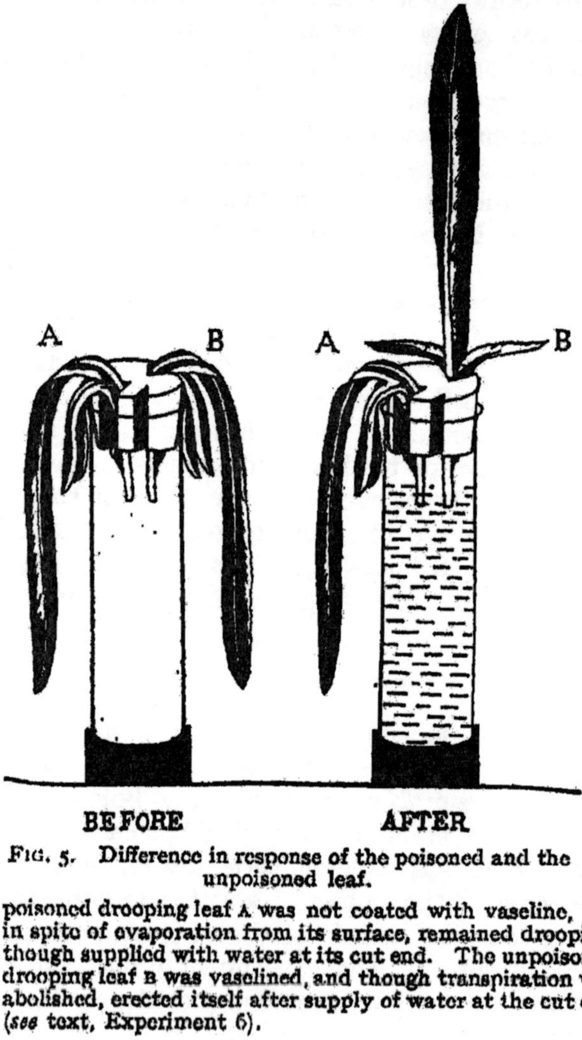

FIG. 5. Difference in response of the poisoned and the unpoisoned leaf.

The poisoned drooping leaf A was not coated with vaseline, but in spite of evaporation from its surface, remained drooping, though supplied with water at its cut end. The unpoisoned drooping leaf B was vaselined, and though transpiration was abolished, erected itself after supply of water at the cut end (*see* text, Experiment 6).

not coated with vaseline, while the living one, B, was coated with it. The two leaves were mounted in the same bottle and their cut ends irrigated afterwards with warm water.

If, as according to the physical theory, transpiration or evaporation were the real motive power for the movement of sap, the sap should rise in the killed uncoated leaf, and the rise would be manifested by an erectile movement. In the vaselined living leaf, on the other hand, there should be no such rise of sap, since transpiration has been abolished. But the experimental results are exactly the opposite of what was to be expected from the physical theory; for while there was not the slightest indication of rise of the killed unvaselined leaf A, the living vaselined leaf exhibited a rapid erectile movement (fig. 5).

Normal and Reversed Direction of Flow of Sap

The statement has already been made that the movement of sap is determined by differential activity at the two ends of the organ, and that the flow of sap follows the turgor-tension gradient from the more to the less activated region. This indicates the possibility of inverting the flow of sap from the normal upward to the downward direction by reversing the turgor-gradient. This has already been demonstrated in my previous experiments carried out with stems. A wider generalisation will now be established by the demonstration of parallel reactions in the midrib of leaves.

Experiment 7.—Two leaves of Cornflower were allowed to droop to a considerable extent. After the removal of the lamina the midribs were coated with vaseline. They were suitably mounted in opposite ways, shown as (*a*) and (*b*) in fig. 6.

Transport in normal direction.—In (*b*) it was the apical end A that hung down, the response being indicated by the movement of an attached index against a scale. Application of water at the freshly cut basal end B caused a pronounced erectile movement of the end A, demonstrating the flow of sap in the normal centrifugal direction. The movement of the index against a scale graduated in degrees gave the rate of erectile movement which indicated the rate of propulsion in the normal centrifugal direction.

The time of erection through 15 degrees was thus found to be 77 seconds, or 5·1 seconds per degree.

Transport in the reverse direction.—In the arrangement (*a*) water was applied to the apical end A: the basal end of the midrib now exhibited erectile movement, proving that the sap was transported in the centripetal direction or against the normal. The rate of erection was 15 degrees in 99 seconds, or 6·6 seconds per degree.

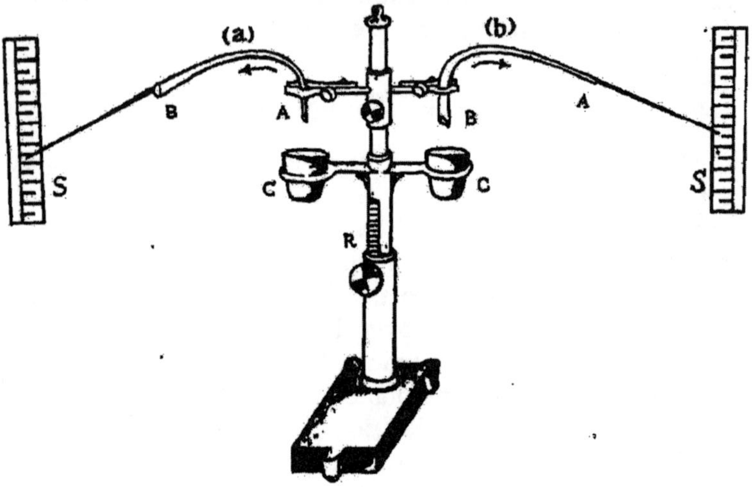

FIG. 6. Transport of sap in vaselined midrib in the centrifugal (*b*) and in the centripetal direction (*a*).

Water is applied from below in cups which can be raised by a rack and pinion. The rate of erectile movement is measured by movement of an attached index against a scale.

It would thus appear from the results of these two experiments that the velocity of transport is not the same in the two directions, but that the normal centrifugal velocity is relatively the greater. This conclusion is supported by the results of other experiments which I had previously carried out.

The important conclusion to be drawn from the above experiments is that the transport of sap occurs in the complete absence of transpiration, and that the direction of propulsion is from the point activated by irrigation to the distant inactive region (*cf.* Experiment 27, p. 54).

Factors modifying Erectile Response

The experiments described prove that the activity of the ascent of sap and its induced variations may be investigated by the method of erectile response of the leaf. This is a direct and simple method free from any complications. It has been shown that there is an independent activity which will be designated as Ai, by which the water absorbed at the point of application is propelled onwards. This pumping up of sap increases the turgor, and making the midrib more rigid brings about an erectile movement. The loss of water, Lt, at the transpiring surface of the leaf induces, on the other hand, an effect precisely the opposite, causing flaccidity of the midrib and a tendency towards the fall of the leaf. The attitude of the leaf would then obviously depend on Ai–Lt. The following are the more important and representative types of reaction; the leaf is supposed to be in a slightly drooping condition to begin with.

(1) Ai > Lt.—The active suction is supposed to be enhanced by a stimulating agent, transpiration remaining normal. Absorption being then greater than loss, the response of the leaf is an erectile movement.

(2) Ai = Lt.—The absorption of water is equal to the loss by transpiration. The gain and the loss being the same, the leaf remains stationary, exhibiting neither a rise nor a fall.

(3) Ai < Lt.—The rate of suction is supposed to be lowered by a depressant, transpiration remaining normal. The absorption is less than the loss, and the resulting response is drooping or fall of the leaf.

Having thus obtained a clear conception of the responsive mechanism of the leaf, attempts were next made to make the leaf a sensitive and accurate indicator of the activity of the ascent of sap and its induced variations.

In regard to the choice of suitable specimens, the more quickly they respond to drought and irrigation, the more useful are they for experimental purposes. The special advantage of active leaves is the reduction of the period of the experiment during which the environmental conditions should remain practically unchanged. No specimens are probably so active as the rosette of long leaflets borne on the petiole of the Lupin. This plant flourishes during summer in different countries of Europe, and the demonstrations with it that I have been called upon to give before English and Continental Universities proved to be invariably successful. Good results with Lupin were also obtained at my Mayapuri Research Station in Darjeeling, situated at a height of 7000 feet, the climatic conditions of the place being similar to those of Europe.

The climate of Bengal is, unfortunately, very unsuitable for the growth of the plant, and it was therefore necessary to discover other plants whose leaves were moderately active and could be substituted for those of Lupin. The characteristics of some of these leaves are given in the next paper.

SUMMARY

The responsive movement of a drooping leaf supplied with water affords a very sensitive and reliable indication of the ascent of sap and its induced variations.

The permanent abolition of the erectile movement by the action of poisonous solutions proves the existence of a vital factor in the transport of sap.

The persistence of the rise of sap in a leaf coated with impermeable varnish proves the existence of an activity independent of transpiration.

The flow of sap is directed from the more to the less active region. The direction of flow can be inverted by reversing the differential activity at the two ends of the organ.

III.—CHARACTERISTICS OF THE ERECTILE RESPONSE OF LEAVES

BY

UMESH CHUNDER SEN, M.SC.

As soon as a leaf is isolated from the plant it begins to droop, and the drooping increases with the duration of isolation. Application of water at the cut end of the leaf not only arrests this down-movement, but reverses it into one of erection, the rate of which depends on the propulsive activity of the absorbing end of the organ. In order to develop this reaction into a method of precision for research, the first condition is to secure fairly active specimens and to become acquainted with their behaviour under different conditions. Reference has already been made to the activity of the leaves of the Cornflower (*Centaurea lanata*) and of *C. moschata*. Though they belong to the same genus, yet their physiological activity exhibits a certain qualitative difference.

Cornflower (*C. lanata*).—This plant grows in Calcutta in winter from December to March. Its leaves, when fully vigorous, are almost as active as the leaflets of Lupin. The lanceolate leaf of this plant has sometimes two lobes or arrow-like projections of the lamina near the base. The weight of the leaf is inconsiderable ; and the erectile movement of the leaf, after irrigation, is therefore not opposed to any great extent by its weight. Most of the experiments, described below, were carried out during the months of December, January, and February.

C. moschata.—The leaf is spatulate in shape and is longer, broader, and thicker than that of the Cornflower.

Its activity, generally speaking, is less pronounced; the weight of the leaf to be lifted during the erectile response is, moreover, greater than in the Cornflower.

The experimental method depends on the determination of the rate of erectile response in a batch of leaves in a uniform condition of drought, the specimens being supposed to be equally active. It is then possible to determine the average rate of the erectile response under normal conditions and the change induced under a definite external variation. In order that these results should be comparable with each other, it is necessary to have means of securing a uniform condition of drought in a batch of equally vigorous specimens.

Production of Uniform Condition of Drought

Considerable difficulty was encountered in inducing this condition. A batch of cut leaves was at first allowed to wither in the experimental room, the condition of drought being judged from the limpness of the leaves. This method of natural desiccation was somewhat uncertain, as certain days were dry and warm, while others were damp and cold. Exposure to the sun caused drought more quickly, but uniformity of sunlight could hardly be depended upon. The difficulties were, however, overcome by a device for securing uniform desiccation in a batch of leaves. Three degrees of desiccation could thus be obtained, representing either (1) feeble, (2) moderate, or (3) excessive drought. In the condition of *feeble* drought the free end of the leaf bends through 90 degrees and assumes an approximately horizontal position; under *moderate* drought the leaf bends down vertically through 180 degrees; a still greater amount of wilting is described as *excessive* drought. The leaf is supposed to bend over with the dorsal surface facing upwards. The contrivance for securing any particular degree of drought is described below.

The Electric Desiccator.—A cone is made of tin, the inner surface of which is highly reflecting. The height of the cone is 33 cm., the diameter of its circular end being also 33 cm. A 220-volt, 32-c.p. incandescent lamp is suitably mounted at the upper end of the cone. A number of similar cut leaves, say, of Cornflower, are spread side by side on a circular piece of white paper, the cone being placed a few mm. above it (fig. 7). The radiation of heat and light within the enclosed space brings about drought in the leaves, the extent of which depends on the duration of exposure.

FIG. 7. The Electric Desiccator.

Radiant heat and light induce drought in the leaves placed in the conical chamber to an extent which depends on the duration of exposure.

The following examples explain the efficacy of this method. When exposed in the conical chamber for a minute and a half, a number of cut leaves of Cornflower exhibited a condition of *feeble* drought, the free end of the leaves bending through 90 degrees from the normal erect position. When the exposure was prolonged to 3 minutes, the condition of *moderate* drought was shown by the bending of the leaves vertically downwards. *Excessive* drought was induced by a still longer exposure. A 16-c.p. electric lamp enables a finer gradation of the extent of desiccation, and this would be preferable, if circumstances did not necessitate the shortening of the period of experiment.

Less active specimens, such as the leaves of *C. moschata*,

require a relatively longer period of exposure for attaining a particular condition of drought; thus *moderate* drought is induced in *C. moschata* after an exposure in the conical chamber for 9 minutes, in place of 3 minutes for the Cornflower.

Securing Equally Active Specimens

Greater difficulties are encountered in obtaining a number of specimens whose activity is approximately equal. This activity is dependent on:

The species of plant.—*C. lanata*, or Cornflower, as previously stated, is usually more active than *C. moschata*, but there are individual exceptions. A particular leaf of *C. moschata* may thus happen to be more active than one of Cornflower.

The season.—The activity of the leaf is found to be considerably modified by this factor. For example, during the dry winter months, December and January, the activity of different batches of leaves was moderate and more or less the same. But in February the spring season was heralded by showers of rain, which lasted for two days, and it was surprising to find how the general activity of the leaves was in consequence increased at least three times. For securing comparative results, each particular series of experiments should be completed in the course of the same day. The best time for experiment is at or about mid-day.

The age of the leaf.—Moderately young leaves at the middle of the stem are found to be more active than the quite young ones at the top and the old ones near the base. Those from the southern aspect, exposed to the sun, are somewhat more active than those from the opposite side.

Bearing in mind the above characteristics, a number of equally active leaves can be selected for securing average results. It has been said that the activity of a leaf may occasionally prove to be very high or very low. Ignoring these exceptional cases, the results given may be regarded as typical.

The average activity of a batch under normal conditions being thus found, it is then possible, through observation of the variation induced in the average rate of erectile response, to determine the effect of environmental change in modifying the propulsive activity.

METHOD OF OBSERVATION

A leaf, say, of Cornflower, is taken in a condition of moderate drought, so that its apex bends vertically downwards. The erect position is marked as 180°, the horizontal as 90°, and the vertically down as the 0° position. The leaf is suitably mounted in the cork of a small bottle; a V-shaped hole allows the introduction of any given liquid into the bottle, until the surface of the liquid comes into contact with the freshly cut lower end of the midrib. If the liquid is actively absorbed and propelled, the effect is shown by the rapid rate of the erectile movement. The observation of this is facilitated by two definitely marked points of reference, namely, that of horizontal or 90° position and the vertical position of 180° (fig. 8). Does the erection of the leaf proceed uniformly, or is there any definite change in the rate?

Time-relations of the erectile movement.—The time required for the erection of the leaf, from the vertically downward 0° to the horizontal position 90°, is first noted and designated by A. The total time B required for erection from the vertically down 0° to the vertically up 180° position is also noted. From A is found the time of erection from 0° to 90°, and from B-A, the time for the further rise from 90° to 180°.

For obtaining the average result, a number of equally active leaves in the same condition of drought are mounted side by side in bottles which are fixed by plasticine in a row on the table. From four to six experiments, with the same species of plant, can thus be carried out at the same time. Parallel experiments, whenever practicable, are also carried out in the same manner with a different species of plant.

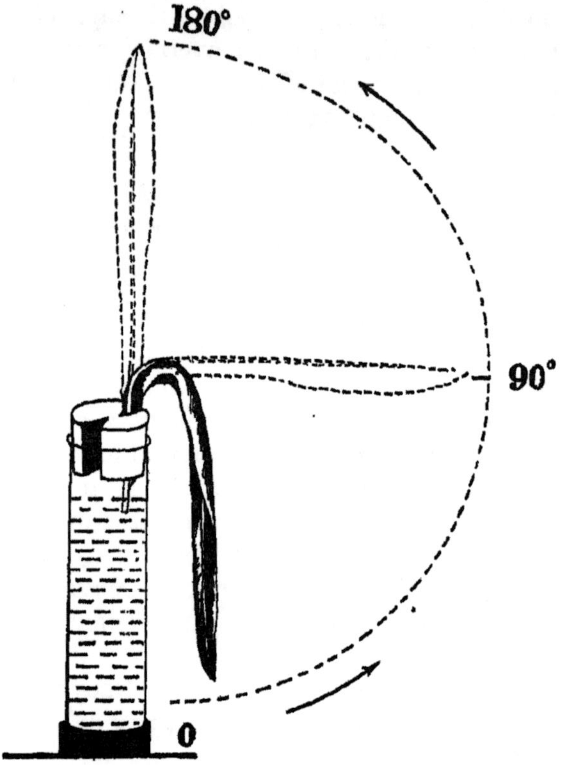

FIG. 8. Method of determination of average rate of erectile response from 0° to 90° and from 90° to 180°.

Leaf under moderate drought droops down vertically to 0°; application of water at the lower end of midrib causes continuous erection to the vertical 180° through the intermediate 90° position.

ERECTILE RESPONSE OF LEAF OF CORNFLOWER

Experiment 8.—The results obtained with the Cornflower will now be given. Simultaneous observations were made of the responsive movement of six leaves in a condition of moderate drought produced by exposure in the conical chamber for 3 minutes. It was a dry day (January 12, 1930), and the temperature of the experimental room was 21° C. Application of water at this temperature to the cut end of the midrib brought about a continuous erection of the leaf from the vertically down to the vertically erect position. The typical results obtained with six different specimens are given in the following table:

TABLE I.—ERECTILE RESPONSE OF LEAF OF CORNFLOWER UNDER MODERATE DROUGHT, AFTER APPLICATION OF WATER AT 21° C.

Number of specimens	Time of erection in minutes						time in minutes
	1	2	3	4	5		
A. Through 90°	3·5	3·5	4·0	4·0	4·0	4·5	
B. ,, 180°	6·7	6·7	6·7	7·0	7·0	7·5	7·0

A = 3·9 minutes; B-A = 3·1 minutes

It will be noted that, individually and collectively, the time required for erection of the leaf from 90° to 180° is a little shorter than from 0° to 90°. Referring to the average value, it is found that it required 3·9 minutes for the leaf to erect itself from 0° to 90°, whilst the time of erection from 90° to 180° (B-A) is a little shorter, being 3·1 minutes.

ERECTILE RESPONSE OF LEAF OF *C. MOSCHATA*

Experiment 9.—Observations were taken, as in the last case, with six different specimens, which were brought to a condition of moderate drought by exposing them for 9 minutes inside the conical desiccator, in the place of 3 minutes for Cornflower. Table II gives the typical results.

TABLE II.—ERECTILE RESPONSE OF *C. moschata* LEAF UNDER MODERATE DROUGHT, AFTER APPLICATION OF WATER AT 21° C.

Number of specimens	Time of erection in minutes						Average time in minutes
	1	2	3	4	5	6	
A. Through 90°	8	8	9	10	12	13	10
B. " 180°	15	15	17	18	20	22	18
	10 minutes ; B–A = 8 minutes						

In *C. moschata* also the rate of erection from 90° to 180° is quicker than that from 0° to 90°.

EFFECT OF EXCESSIVE DROUGHT

Experiment 10.—Excessive drought was induced in a batch of leaves of Cornflower by an exposure of 6 minutes within the conical chamber. The following results are typical of the effect of excessive drought on the rate of erectile movement:

TABLE III.—ERECTILE RESPONSE OF LEAF OF CORNFLOWER UNDER EXCESSIVE DROUGHT, AFTER APPLICATION OF WATER AT 21° C.

Number of specimens	Time of erection in minutes						Average time in minutes
	1	2	3	4	5	6	
Through 180° .	22	25	28	35	26	32	28

It will be noted that while under moderate drought the average time was 7 minutes for erection from 0° to 180°

(*cf.* Table I), this was prolonged under excessive drought to 28 minutes. Excessive drought has thus the effect of slowing down the rate of erectile response, presumably due to a slower rate of ascent of sap.

Enhancement of Activity in Spring

Experiment 11.—As an example of the efficiency of the method of erectile response for the detection of the invisible internal change of physiological activity, mention may be made of the sudden increase of the rate of erectile response which occurred in early spring (February), after a shower of rain. This increase denoted the enhancement of activity in the propulsion of sap. The following table gives the typical results:

Table IV.—Enhancement of Activity in Spring. Leaf of Cornflower under Moderate Drought, after Application of Water at 21° C.

Number of specimens	Time of erection in minutes						Average time in minutes
	1	2	3	4	5	6	
Through 180° .	1·3	2	2	2·2	2·2	2·4	2·0

It will be remembered that the average time of erection of the leaf of the Cornflower through 180°, a few days before, was 7 minutes (*cf.* Table I); but the favourable season caused a sudden increase of activity, which was about three and a half times as great. In some instances the drooped leaf after irrigation sprang up with incredible rapidity, leaving no doubt that a very active force was causing propulsion of sap.

SUMMARY

Leaves which have drooped vertically downwards under drought re-erect themselves vertically after supply of water to their cut ends.

Different degrees of drought can be produced by the Electric Desiccator.

The rate of erectile movement after irrigation undergoes a continuous increase. The erection from 90° to 180° is quicker than from 0° to 90°.

The time of erectile recovery is greatly prolonged in leaves which have been subjected to excessive drought.

The movements observed by the method of erectile response are, clearly, the expression of variations in the rate of movement of the sap induced by changes in the external conditions which affected the physiological activity of the leaf.

IV.—METHODS OF RECORD OF THE ERECTILE RESPONSE OF THE LEAF BY THE ELECTRO-MAGNETIC TAPPER AND THE AUTOMATIC RECORDER

BY

SIR J. C. BOSE

It has been explained in the last paper how the average activity of a number of leaves can be determined, and how the variation induced in the average activity affords an indication of the effect of change of the environment on the propulsion of sap. The ideal method would, however, be one by which the effect of physiological variation would be demonstrated in one and the same specimen. The problem was solved by the invention of a new and very delicate method.

I find that a very sensitive condition of the leaf for the manifestation of any change of activity is obtained when it is in a state of feeble drought, the drooping leaf being horizontal, or at an inclination of 90 degrees or less from the vertical. The slightest change in the activity of transport of sap is then shown by a resulting modification of the rate of the erectile movement. Thus any enhancement of activity in the propulsion of sap is exhibited by an erectile response more rapid than the normal rate, while depression is shown by the slowing down of the rate, culminating in an arrest or even an actual reversal of the movement from positive to negative according to the effectiveness of the depressant. The opposed responsive movements may be appropriately indicated by the signs positive and negative: the positive (+) indicates the erectile response, while the negative (—) stands for the opposite movement of fall or drooping of the leaf.

In the following investigations it is, however, first necessary to find satisfactory means of mounting the leaf in a carrier and of subjecting the cut end of the leaf alternately to different solutions which cause either stimulation or depression; secondly, suitable device has to be made for

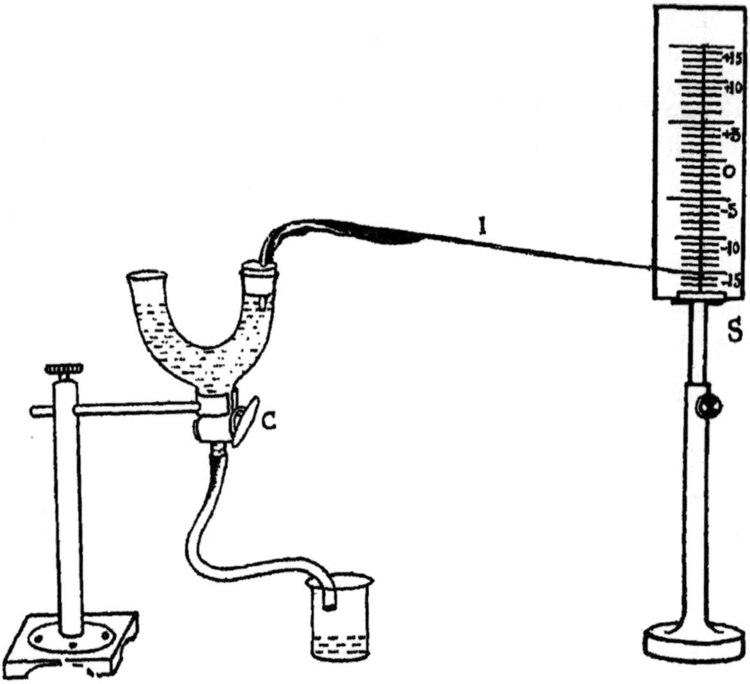

FIG. 9. The Indicator of the movement of leaf.

I, index moving in front of scale S on mirror. Different solutions can be introduced to cut end of midrib and withdrawn by manipulation of stopcock C.

magnification of the movement of the leaf and for the measurement of the exact rate.

The Carrier.—Several types of carrier have been devised to serve different purposes; the first of these consists of a short U-tube provided with a stopcock. The drooping leaf is suitably mounted by means of a cork on the right limb of the tube, the end of the other limb being open for the introduction of water or other solution, into which the end of the midrib of the leaf projects below the cork. The leaf continues to droop as long as there is no supply of water; but when water is supplied through the free left end of the tube, the movement of fall becomes arrested, and an erectile movement of the leaf is initiated by the pumping up of water absorbed at its cut end. Different chemical solutions can be quickly substituted for water by proper manipulation of the stopcock C (fig. 9). For the study of the effect of water at different temperatures, the precaution is taken of enclosing the U-tube in a cover of non-conducting material, so that a definite temperature can be maintained during the course of the experiment.

A different method of mounting the drooping specimen is to hold it by a clip about 3 cm. above the cut end of the midrib of the leaf. Cold or hot water in small cups may then be applied from below. The same procedure is adopted for applying different chemical solutions. The cups can be rapidly raised or lowered by a rack and pinion (*cf.* fig. 6).

It is next necessary to magnify the movement of the leaf by an attached indicator, so that the normal rate of movement and its induced variation may be measured.

THE MAGNIFYING INDICATOR

The leaf is mounted under feeble drought, which is more satisfactorily obtained by allowing the leaf to droop naturally after isolation rather than subjecting it to more pronounced desiccation in the conical chamber. An exceedingly fine glass thread, with practically no weight, is then attached to the partially drooping leaf by a drop of shellac varnish. The total length of the indicator (including a short length of the leaf) is 18·2 cm.

The Scale.—The index moves in front of a scale engraved on a strip of glass mirror, the successive marks being 3 mm. apart. A movement of the index of length 18·2 cm. through a distance of 3 mm. then indicates a rise (or a fall) through 1 degree. The scale on the mirror has the advantage that the error of parallax is avoided when the eye of the observer, the end of the index and its reflection in the scale are in the same line (*cf.* fig. 9). A still greater accuracy of observation is secured by interposing a magnifying lens between the eye and the index.

The time required for the index to rise through successive degrees in the scale after irrigation of the leaf can be noted with the help of a metronome which beats, say, once in a second. The measurement of the time for the rise of the index through successive degrees requires the co-operation of two observers, one signalling the time of transit of the index through successive degree-marks on the scale, and the other noting the time-interval from the ticks of the metronome. In this way the time required for the index to rise through successive single degrees of the scale (the first, the second, the third degree and so on) is obtained. The object of this procedure is to ascertain whether the rate of erectile movement under normal conditions remains fairly constant.

The employment of two observers with different personal equations may possibly lead to some small error in the measurement of time. I therefore devised a method which enables one single observer to obtain a direct record of the time-intervals. This was secured by the Electro-magnetic Tapper.

Record on Revolving Drum by Electro-magnetic Tapper

The Drum.—A carefully turned cylindrical drum of aluminium is revolved at a uniform speed by clockwork; this is secured by a perfect governor, by which the rate of revolution can be suitably increased or decreased, or kept perfectly uniform. In practice two definite speeds are found most useful: one in which the recording surface moves through 1 mm., and the other through 3 mm. per second. The Electro-magnetic Tapper which inscribes the record can be suitably moved by a rack and pinion, so that two or three series, taken one above the other, under different conditions can be compared.

The Electro-magnetic Tapper.—This consists of a pair

of small electro-magnets activated by an electric current, which is completed for a moment by a press-key, K. The spring which carries the soft iron armature is prolonged for the attachment of a writer, which marks successive dots

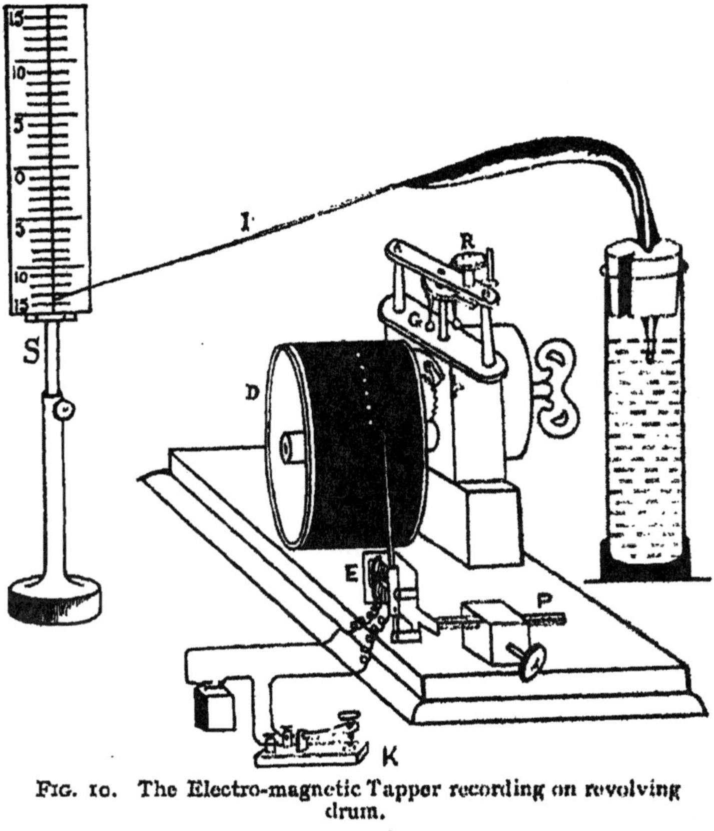

FIG. 10. The Electro-magnetic Tapper recording on revolving drum.

Index I attached to leaf moves in front of scale S. Pressure of key K actuates electro-magnet E and inscribes dot on smoked paper on surface of drum D, whose speed is adjusted by regulator R. Rack and pinion P moves the tapper for series of successive records.

on the smoked surface of the recording paper. Or the dots may be inscribed in black on white by means of a small stylographic pen recording on ordinary paper (fig. 10).

The method of taking a record is as follows: The drum is released by pulling a catch, so that the rate of its revolution soon becomes uniform. The leaf mounted in the carrier is first allowed to droop till the index falls, say, about 20 degrees below the horizon. Water, as previously explained, is next supplied to the cut end of the leaf, with resulting erectile movement which becomes steady by the time the index reaches the line marked $-15°$. The observer now presses the key and a dot is marked on the recording paper. This procedure is continued for recording the movement of the index through rise of every degree from -15 to $+15$. This is the median range of erection, of which one-half is below and the other above the horizon.

The record permits the determination of successive periods of erectile response per degree. The intervals of time can be easily measured from the spacings between successive dots; in the quicker-moving drum the speed of the recording paper is 3 mm., while in the drum moving at a slower speed it is 1 mm. per second. The record obtained is sometimes too long for printing within the breadth of a page. In that case it has to be suitably reduced by photo-mechanical process. This, however, does not stand in the way of absolute determination of the rate of erectile movement, for a definite time-interval is inscribed below the record. In other cases the time for the erection of the leaf per degree of space is mentioned in the explanatory description of each record.

The Rate of Normal Erectile Movement

In the widest range of erection of the leaf, the rate of rise was found to undergo a slow and continuous increase from vertically downward to the vertically upward position (*cf.* Table I).

In the method of the Electro-magnetic Tapper the record of erectile movement is taken through the very much shorter range of 30°, from 15° below to 15° above the zero or horizontal position. The erectile movement within this narrow range increases at such a small rate that it may be regarded as practically uniform, as will be seen from the following records.

Experiment 12. *Record of Cornflower.*—Two series of records were obtained with a moderately active leaf of Cornflower, the first during its rise from −15° to 0°, and the second from 0° to +15°. The successive dots, it will be remembered, are inscribed at intervals corresponding to the rise of the leaf through 1 degree of space. The line inscribed below represents a period of time of 40 seconds, during which the leaf erected itself from −15° to 0°, the average rate of erection for the 15 degrees being 2·7 seconds per degree. The lower row of the record gives the time of erection from 0° to +15°, the average rate of which is 2·4 seconds per degree, and, though slightly quicker than that in the first row, is more or less of the same order.

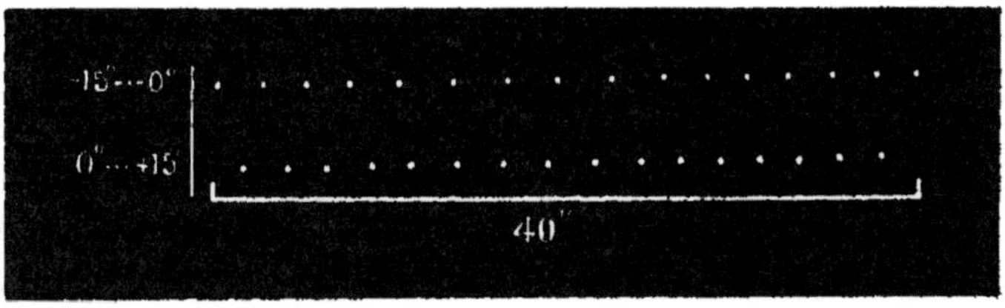

FIG. 11. Records giving rate of erectile movement of leaf per degree from −15° to 0°, and from 0° to +15° (Cornflower).
(The horizontal line below represents a period of 40 seconds.)

In regard to the rate of erectile response of Cornflower throughout every degree, the values calculated from the record (fig. 11) are given in Table V. Though the time for erection per degree is very nearly the same throughout the range, yet careful examination detects a periodic variation, seemingly of a pulsatory character.

In Experiment 12 the leaf of Cornflower employed was moderately active, average rate of its rise being 2·5 seconds per degree. In less active leaves the rate is half as slow, as was found in a second record, in which the leaf rose from −15° to 0° in 82 seconds, and from 0° to +15° in 67·5 seconds. The average rate in the two cases is 5·4 and 4·5 seconds per degree respectively. A still slower rate is given by an older leaf.

TABLE V.—TIME-INTERVALS FOR ERECTION PER DEGREE, FROM − 15° TO + 15° (CORNFLOWER).

	Degrees		Degrees	Seconds		Degrees		Degrees	Seconds
From	− 15	to	− 14	2·6	From	− 0	to	+ 1	2·6
,,	− 14	,,	− 13	2·6	,,	+ 1	,,	+ 2	2·3
,,	− 13	,,	− 12	2·6	,,	+ 2	,,	+ 3	2·6
,,	− 12	,,	− 11	3·0	,,	+ 3	,,	+ 4	2·3
,,	− 11	,,	− 10	3·0	,,	+ 4	,,	+ 5	2·6
,,	− 10	,,	− 9	3·0	,,	+ 5	,,	+ 6	2·6
,,	− 9	,,	− 8	3·0	,,	+ 6	,,	+ 7	2·6
,,	− 8	,,	− 7	3·0	,,	+ 7	,,	+ 8	2·6
,,	− 7	,,	− 6	3·0	,,	+ 8	,,	+ 9	2·6
,,	− 6	,,	− 5	2·6	,,	+ 9	,,	+ 10	2·3
,,	− 5	,,	− 4	2·3	,,	+ 10	,,	+ 11	2·3
,,	− 4	,,	− 3	2·3	,,	+ 11	,,	+ 12	2·3
,,	− 3	,,	− 2	2·3	,,	+ 12	,,	+ 13	2·3
,,	− 2	,,	− 1	2·6	,,	+ 13	,,	+ 14	2·3
,,	− 1	,,	− 0	2·3	,,	+ 14	,,	+ 15	2·3

Experiment 13. *Record of C. moschata.*—This is given in fig. 12; the results are similar to those with the

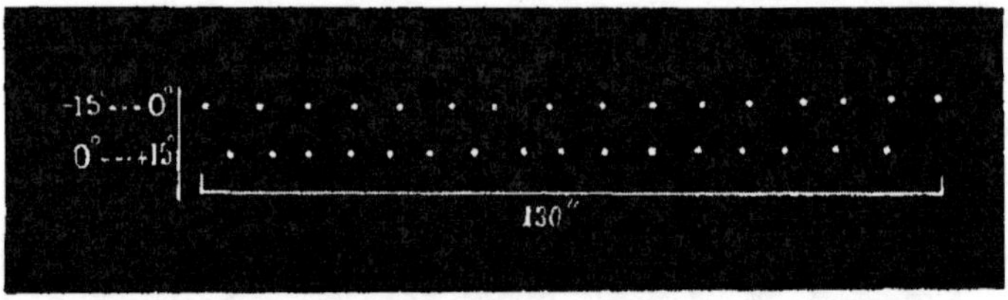

FIG. 12. Record of erectile movement of leaf of *C. moschata* for every degree of rise from − 15° through 0° to + 15°.

Cornflower, the only difference being that the rate of erectile movement is relatively sluggish. The average rate from − 15° to 0° is 8·6 seconds, whereas from 0° to + 15° it is 8·1 seconds, per degree.

In regard to the choice of species of plant for the satisfactory demonstration of the variation induced by physiological change, I find that the employment of Cornflower has many advantages: first, the leaves are generally so active that the experiments carried out with them have invariably proved to be successful; secondly, they quickly exhibit reaction to physiological variation, so that the characteristic effects are exhibited within the short time available for the record during the rise of the leaf from $-15°$ to $+15°$. It is necessary that the experiment should be completed within a short time, since no record is possible when the leaf has erected itself nearly to the vertical. *C. moschata*, being more sluggish, takes a longer time for exhibition of the characteristic reaction.

The tapping record gives a strikingly visible demonstration of the change of physiological activity induced by external agents. When the activity is uniform, the successive spacings between the dots are approximately equal. A depressing agent causes an abrupt change, shown by the wider separation between the successive dots. A closer approximation of the successive dots, on the other hand, is the characteristic effect of a stimulating agent.

Response by Drooping and by Erection

It has been stated that the method of recording the erectile response by the Electro-magnetic Tapper is extremely sensitive for detection of any variation in the rate of ascent of sap. This rate is practically uniform under normal conditions when water is being absorbed at a constant rate at the cut end of the leaf. If, however, the power of absorption and propulsion is interfered with in any way, then the resulting erection of the leaf will undergo a decline. As an extreme example, the case may be considered when both absorption and propulsion become zero by the withholding of water. During the withholding there is a loss of water by transpiration from the surface of the leaf which causes increasing drought and limpness of the midrib, exhibited by drooping or negative response. Subsequent irrigation causes, however, a revival of propulsive activity, which results in the erectile or positive response of the leaf. These characteristic effects are demonstrated in the following experiments.

Experiment 14. *Negative and positive response.*—A pair of records was obtained with an identical leaf of Cornflower.

The upper record D shows the rate of drooping or negative response, which was 8·3 seconds per degree. After application of water the negative response of fall was, after a short while, transformed into the positive erectile

FIG. 13. Negative response caused by drought transformed into positive after application of water.

D — is the negative response; W + is the positive after application of water.

response, the rate of which was 3·5 seconds per degree (fig. 13). In a second pair of records the rates of drooping and of erection were found to be nearly equal, being respectively 13·3 and 13·4 seconds per degree. The difference is due to the fact that the rates of absorption and of transpiration were not the same in the two cases.

THE PERIOD OF TRANSITION

The negative drooping response cannot, however, be at once transformed into the erectile positive without what may be described as a period of negotiation. There are evidently two opposed forces at work: the one is the loss of water from the transpiring leaf, and the other is the accession of water due to the propulsive activity initiated by irrigation of the cut end of the leaf. During this important period of transition, it would seem almost impossible to detect what is happening, when the two opposing forces are at work. A continuous record throws much light on this obscure point.

Experiment 15. *Record of the period of transition.*—It is difficult to reproduce, in a limited space, the entire record with its changes from negative to positive through the intermediate phases of transition. The record (fig. 14) is in three rows: the first exhibits the rate of fall during withholding of water; the second is that of the transition period immediately after the supply of water, when the rate of drooping or negative response becomes less and less negative, till it is transformed into positive as shown in the third row. Detailed measurements show that the rate of negative response by fall D —, during withdrawal of water,

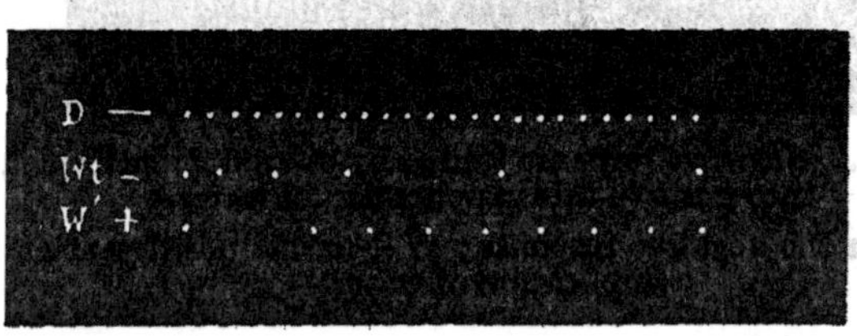

FIG. 14. Record of period of transition from negative to positive.

Note in the second row the gradual slowing down of the negative after application of water, and the gradual quickening of rate of positive erectile response in the third row.

was 3·3 seconds per degree; supply of water caused increasing delay in the rate of fall Wt —, and the maximum delay of 32 seconds per degree was attained in the course of 83 seconds; the response was then transformed into positive W′ +. The feeble rate of positive response at the beginning, 21 seconds per degree, was increased in a short time to the steady rate of 9 seconds per degree.

In regard to the time consumed in the transformation of the negative into positive, the variation in this transition period gives a very interesting indication of the propulsive activity of the specimen. Everything else being the same, any enhancement of the propulsive activity shortens the period of transition.

THE AUTOMATIC METHOD OF RECORD

Another important method, illustrated in fig. 15, is the one in which the leaf automatically records its responsive movement. The partially drooped leaf is suitably supported by a clamp C, without any undue pressure.

Arrangement is made, by means of the U-tube previously described, for application of water or other solutions to the lower projecting end of the midrib. The leaf is attached by a cocoon thread to the writing lever W, the

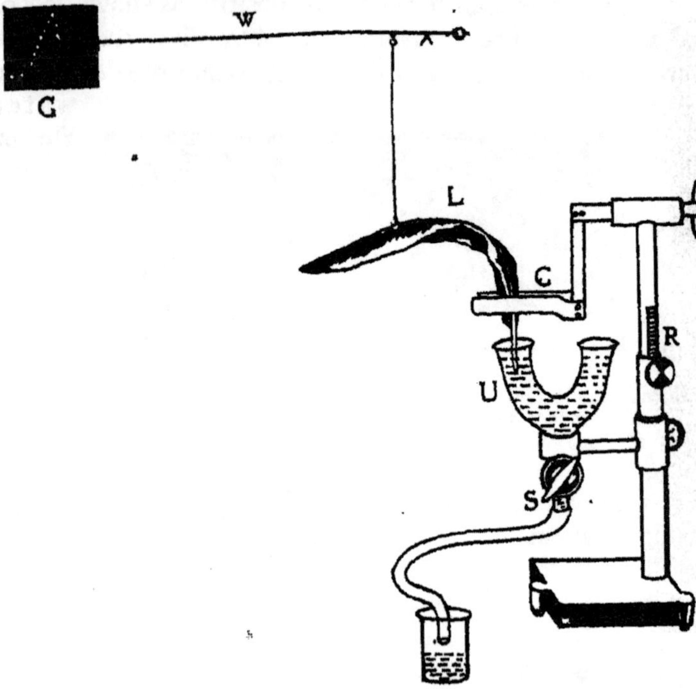

FIG. 15. The method of automatic record.

Drooping leaf L held in a clamp C attached by cocoon thread to writing lever W, which inscribes the responsive movement on smoked-glass plate G kept oscillating by clockwork. Position of the writer W adjusted by rack and pinion R. The U-tube U and stopcock S permits application of different solutions to cut end of leaf.

other arm of the lever being provided with a light counterpoise. The writer is made of a light glass fibre, the tip of which is slightly bent. Special care is taken that the tension exerted on the leaf is so slight as to be negligible. A magnification of 5 to 10 times is sufficient for most purposes. The record is taken on a moving piece of smoked glass, which is kept oscillating backwards and forwards at intervals of 4 or 8 seconds. The dotted curve gives the time-relations of the movement of the leaf. As an example of the sensitiveness of the method, a record is given (fig. 16) of the effects of successive application and withholding of water to the cut end of the leaf. The propulsive activity initiated by the application of water caused the erectile or positive movement, inscribed as an up-curve; withholding of water, on the other hand, caused a negative response of drooping. The rate of erectile response indicates the activity of ascent of sap; arrest of ascent is shown by the resulting negative response of fall of the leaf.

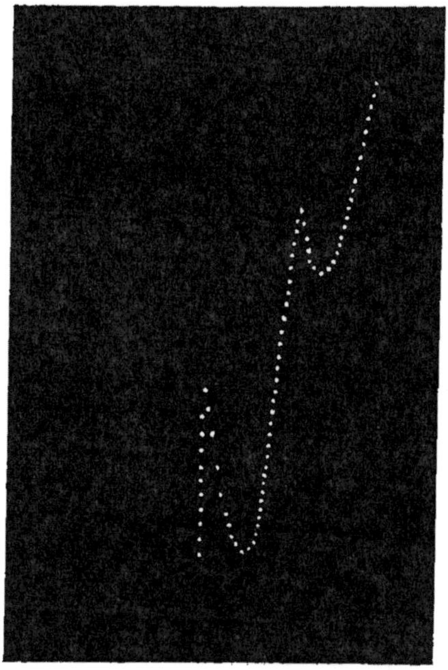

FIG. 16. Automatic record of effect of supply and withholding of water.
Up-curve, erectile positive response; down-curve, drooping negative re- ive dots at intervals of

Several simple and direct methods which can be rendered highly sensitive are thus available for investigations on the ascent of sap. Each of these afford independent evidence of the physiological activity which underlies the propulsion of sap. These methods, described in subsequent papers, are the following:

(1) *The Method of Averages*, in which the average rate of erectile response of a number of similar specimens is observed, as also the change in that rate under physiological variation.

(2) *The Electro-magnetic Tapping Method.*

(3) *The Automatic Method of Record.*

The Method of the Potograph, which I have previously employed and have now greatly extended, will be found to verify fully the results obtained by the other methods.

Summary

A slightly drooping leaf, bent at an angle of — 15 degrees below the horizon, erects itself after irrigation through zero to + 15 degrees at a rate which is approximately uniform.

The rate of erection per degree is accurately inscribed on the recording surface of a revolving drum by means of an Electro-magnetic Tapper. In a moderately active Cornflower the average time of erection for rise of 1 degree is about 5 seconds. In the sluggish leaf of *C. moschata* the average rate is slower, being about 8·3 seconds per degree.

On withholding water, the leaf exhibits a fall or negative response. This is transformed into positive erectile on supply of water.

The record of the transitional period obtained by the Tapping Method exhibits the action of two opposing forces. The variation in the transitional period gives some indication of the propulsive activity of the specimen.

The Automatic Method gives an independent means of recording the activity of the ascent of sap and its induced variations.

V.—EFFECT OF POISON ON THE ASCENT OF SAP

BY

GURU PRASANNA DAS, L.M.S.

IN regard to the action of poisonous solutions it is clear that, on the physical theory, they cannot modify the power of suction of the plant and the rate of its ascent of sap. If they do so, then the phenomenon cannot be entirely physical, but must be due to the physiological activity of living cells. The fact that a poison causes an irreversible change associated with death has been demonstrated by Experiment 3 (p. 12), in which the plant exhibited a complete collapse from which there was no subsequent recovery. The normal rate of ascent of sap will now be shown to be also profoundly depressed by the action of poison, culminating in total arrest.

It has been explained that the rate of the erectile movement of the leaf gives a reliable indication of the rate of transport of sap. The responsive movement of the leaf can therefore be utilised to demonstrate the physiological reaction to poisonous agents.

In the following investigations several independent methods have been employed, namely:

(1) The Method of Eye Observation;
(2) The Method of Tapping Record;
(3) The Method of Automatic Record; and
(4) The Method of the Potograph.

THE METHOD OF EYE OBSERVATION

The following experiments were carried out with a number of similar leaves of the Cornflower, the reactions of which are more rapid than those of *Centaurea moschata*.

The leaves of this plant began to droop under natural drought as soon as they were isolated from the parent plant. They drooped even more rapidly when their cut ends were placed in a 5 per cent. poisonous solution of potassium cyanide. Similar drooping of the leaf also occurred under the plasmolytic action of a 5 per cent. solution of KNO_3. How would it be possible to distinguish between the two reactions? The difference is to be found in their after-effects.

Experiment 16. *After-effect of* KCN *solution.*—After the cut ends of the leaves were placed in the KCN solution they exhibited complete collapse. There was no recovery after application of fresh water, showing that the leaves had been killed by the poison. This irreversible change may occur in as short a time as 10 minutes or so.

Experiment 17. *After-effect of* KNO_3 *solution.*—The fall of the leaf even under prolonged application of KNO_3 solution is not an irreversible process. Substitution of fresh water revives the leaf, which becomes fully erected vertically in the course of several minutes, as demonstrated in the next experiment.

The Method of the Electro-magnetic Tapper

The difference between the plasmolytic and the poisonous reactions is clearly demonstrated in the two following experiments, the records of which were taken by the Tapping Recorder.

Experiment 18. *Effect of* KNO_3 *solution.*—A leaf of Cornflower was transferred quickly into KNO_3 solution, immediately after it had been cut from the parent plant. A record was then taken of the rate of responsive fall per degree under the action of the plasmolytic solution. On substitution of water the negative response of fall became gradually arrested and ultimately transformed into erectile positive response.

The Transitional Period.—These characteristic variations, as well as the duration of the transitional period, are very clearly seen in fig. 17. The first row exhibits the negative response of fall under KNO_3 solution, the rate of which is 4 seconds per degree. The second row, Wt —, shows that 110 seconds is the transitional period of transformation from negative to positive. The rate of negative

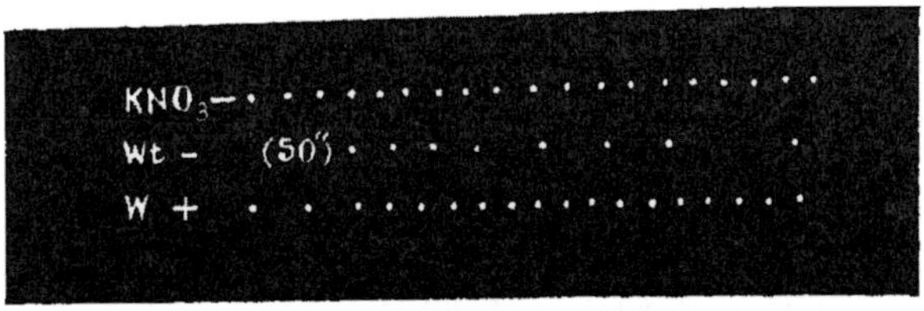

FIG. 17. Effect of application of KNO_3 solution.

The first row indicates negative response of drooping. The second row, Wt —, shows the diminution of the rate of negative response after application of fresh water. Transitional period of transformation from negative to positive is 106 seconds. The third row, W +, exhibits revived positive rate which is gradually increased from 6 to 3·6 seconds per degree (Cornflower).

fall became gradually slowed down in the second stage; a portion of the record during 50 seconds is omitted and indicated by (50″). The rate of fall gradually became slowed down to 16 seconds per degree. A transformation

FIG. 18. Effect of application of KCN solution.

There is no revival from the negative response of drooping even after application of fresh water, as shown in record given in the second row (Cornflower).

then occurred from negative fall to positive erectile response, shown in the third row, marked W +. The rate of this positive response was gradually increased per degree from 6 seconds to the uniform rate of 3·6 seconds.

Experiment 19. *Effect of* KCN *solution.*—A parallel experiment, under the action of poisonous KCN solution, was carried out with a leaf of Cornflower. The leaf underwent a rapid fall, the negative response being seen in the first row. Application of fresh water did not produce any revival as in the last case, but the leaf continued to fall till it collapsed (fig. 18).

The Method of Automatic Record

This method also gave results which are essentially similar to those of the Tapping Method.

The effect of KNO_3 solution is correspondingly modified on varying the duration of application. The two following experiments describe the effects of short and long application.

Experiment 20. *Effect on erectile response of brief application of* KNO_3 *solution.*—A drooping leaf of Cornflower exhibited a more or less uniform erectile positive response after supply of water at its cut end as shown in the up-record (fig. 19). On substitution of KNO_3 solution, the response was transformed into negative in the course of 4 seconds, as shown by the down-curve. The duration of application was relatively short, about 3 minutes, after which substitution of fresh water brought about an erectile recovery. The transformation of the negative into positive response was initiated after an interval of about 16 seconds; this transitional period is shorter than that after a longer application of the KNO_3 solution, as demonstrated in the next experiment.

Experiment 21. *Effect of longer application of* KNO_3 *solution.*—The cut end of the leaf of Cornflower was immediately put into KNO_3 solution and the negative response of fall of the leaf recorded. Water was substituted after 8 minutes (a portion of the record is omitted), and the fall still persisted for about 104 seconds. After this the response of the leaf became erectile or positive. The transitional period for the transformation of the negative into positive was thus 104 seconds (fig. 20), which was practically the same as that obtained by the Tapping Method (*cf.* Experiment 18).

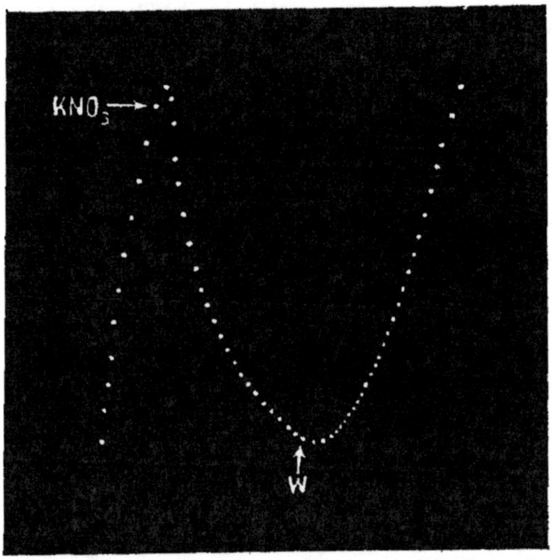

FIG. 19. Record of brief application of KNO_3 solution and substitution of water, by the Automatic Method.

First part of the record shows erectile response of leaf under drought to application of water. KNO_3 solution caused a negative response (down-curve) in 4 seconds. Substitution of water at W transformed negative into positive in the course of 16 seconds (Cornflower). Successive dots at intervals of 8 seconds.

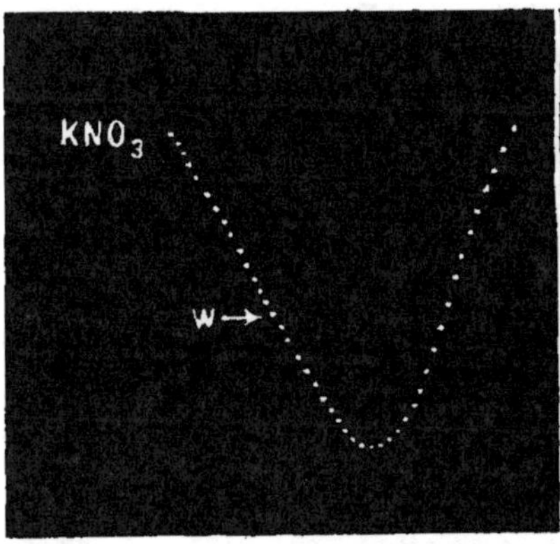

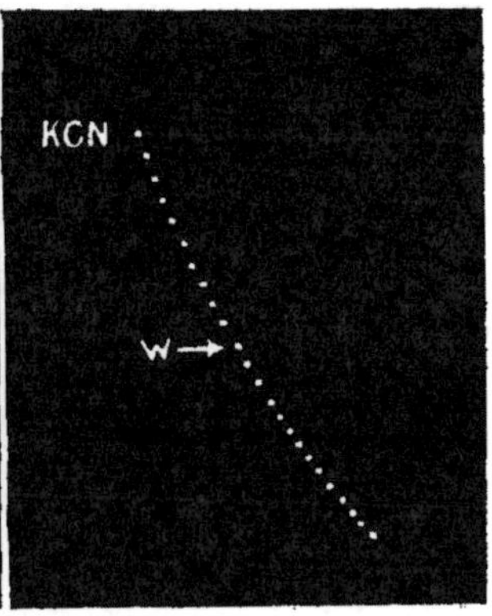

FIG. 20. Effect of longer application of KNO_3 solution, and of substitution of fresh water.

Negative response of fall transformed into positive erectile response after a transitional period of 104 seconds (Cornflower).

FIG. 21 (Right). Effect of KCN solution.

The negative response of fall is not transformed into positive on substitution of fresh water.

Experiment 22. *Effect of* KCN *solution.*—The cut end of the leaf was, as in the last case, immediately placed in the KCN solution and kept in it for about 15 minutes. The first part of the record (fig. 21) is negative. This negative response of fall of the leaf continued even after the substitution of fresh water, and the leaf did not exhibit any recovery. The after-effects of KNO_3 and KCN solutions are thus characteristically different.

The Method of the Potograph

A full description of the Potograph will be found elsewhere.[1] The following short account will facilitate a clear understanding of certain results given in this paper. The apparatus consists of a vessel and attached capillary potometric tube with a contrivance for recording the excursion of a water-index and its time-relations. The description of the recording arrangement is not necessary for the present purpose: the rate of suction is measured by noting the number of scale-divisions per minute through which the water-index moves in the potometric tube (fig. 22) at definite intervals of time. The absolute rate of suction in cubic millimetres is found by multiplying the length of excursion of the index per unit of time by the capillary constant of the tube. Special care is taken to prevent all leakage by smearing the cork holding the leaf with cocoa-butter. The leaves of both the Cornflower and *C. moschata* have been employed for the following investigations:

(1) The effect of electric shock on the normal rate of suction.
(2) The characteristic variation of rate of suction exhibited by a drooping leaf.
(3) The persistence of suction in a vaselined leaf.
(4) Direct and after-effect of a plasmolytic solution on the rate of suction.
(5) The effect of poison on the ascent of sap.

[1] Bose, *Motor Mechanism of Plants* (1928), p. 338.

It should be borne in mind that what is actually observed by the potometer is the absorption of water by the leaf, which is correlated with the ascent of sap to the transpiring surface. The experiments were carried out under two different conditions: first, when the leaf was fully turgid and erect; and, second, when in a condition of drought. In

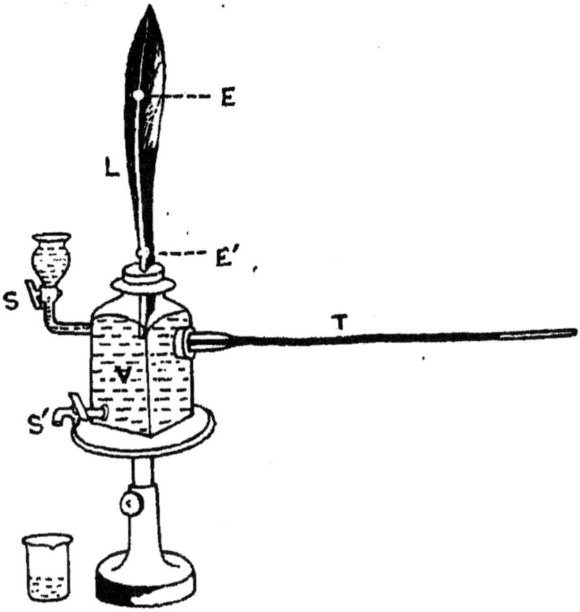

FIG. 22. The Potograph.

S and S′, stopcocks for introduction and removal of solutions; T, capillary tube with water-index; the lower end leaf L, mounted in the cork, dips into the solution. E, E′, electrodes for passage of induction shocks along the midrib.

the first case the normal rate of suction, as measured by the potometer, is constant. The transpiration by the leaves induces incipient drought at the upper end, in consequence of which the sap-stream is propelled upwards at a corresponding uniform rate. This condition may be regarded as one of dynamic balance. If any change of conditions disturbs this balance, it will involve a corresponding variation in the rate of ascent of the sap.

Effect of Electric Stimulation on the Rate of Ascent

Decisive evidence as to the nature of the ascent of sap, whether it is a physical or a physiological process, is afforded by the effect of stimulation on the rate of the ascent. If the phenomenon be purely physical, then electric stimulation will have no effect on the rate of ascent. But if, on the other hand, the propulsion of sap be essentially physiological, then the rate will exhibit a change under electric stimulation. It may be said that, in general, all rhythmic activities are retarded or arrested by sufficiently strong electric stimulation, and that they become revived after a period of rest. As an example of this may be mentioned the arrest of pulsation of the Desmodium leaflet under strong electric stimulation, a gradual revival taking place after the cessation of stimulation. If the ascent of sap be due to rhythmic physiological activity, then under electric stimulation its rate should undergo a retardation, with a subsequent recovery on its cessation.

Experiment 23.—The midrib being the principal channel in the leaf for the transport of sap, electric connections were made with two points on the midrib (*cf.* fig. 22). The normal rate of suction in the leaf of *Centaurea moschata* was 3·7 cub. mm. per minute. The midrib was next subjected for 20 seconds to electric shocks from an induction coil, the intensity of the shock being 3·5 units. The rate of ascent of sap, as measured by the suction, was immediately depressed to 2·1 cub. mm. per minute, the maximum depression of the rate of 1·1 cub. mm. per minute occurring in the course of about 3 minutes. *A very marked retardation of the rate thus took place under electrical stimulation.* The original rate was restored 50 minutes after cessation of stimulation.

Experiment 24.—In a second specimen of Centaurea the rate of suction was 3·3 cub. mm. per minute; the duration of electric stimulation was prolonged to 30 seconds, the intensity remaining the same. The rate of suction was immediately depressed to 2·2 cub. mm. per minute, the maximum depression of 1·4 cub. mm. per minute occurring a few minutes later. There was a complete recovery in the course of an hour.

These very definite effects induced by electrical stimulation offer very strong evidence that the ascent of sap is effected by a physiological mechanism.

Having ascertained the effect of electrical stimulation on suction, observations were made of the variations of the rate of suction in response to changes of environment which would have a physiological rather than a purely physical effect.

CHANGES OF THE RATE OF SUCTION

Experiment 25. *Variation of the rate of suction exhibited by a drooping leaf.*—In this and in the two following experiments the specimens were taken in a condition of drought; after mounting on the Potograph, water at the temperature of the room (20° C.) was applied at the cut end. Observations of the rate were next made at definite intervals. The rate of suction was then found not to be uniform, but to exhibit a very characteristic variation. The following table gives results which are typical:

TABLE VI.—RATES OF SUCTION OF LEAF OF *CENTAUREA MOSCHATA* PREVIOUSLY SUBJECTED TO DROUGHT.

Time-interval after commencement	Rate of suction in cub. mm. per minute
Immediately after	3·4
5 minutes after	4·3
7 ,, ,,	3·1
10 ,, ,,	2·3
15 ,, ,,	2·0
20 ,, ,,	1·8
25 ,, ,,	1·7
30 ,, ,,	1·6
35 ,, ,,	1·6
40 ,, ,,	1·6

The rate of suction at the beginning was 3·4 cub. mm. per minute. This rose to 4·3 cub. mm. per minute, after which it declined continuously, till after the thirtieth minute the uniform rate of 1·6 cub. mm. was attained.

Experiment 26.—Exactly similar results were obtained with the Cornflower. The immediate rate was 2·7 cub. mm. per minute; this was increased to 3·8 cub. mm. in the course of 7 minutes. After this, there was a continuous decrease till the final uniform rate of 1·4 cub. mm. was attained after 40 minutes.

The detailed results given in Table VI clearly indicate three different phases of reaction:

First phase: The rate exhibits an immediate increase, which lasts for a certain length of time.

Second phase: There is a continuous decrease till a uniform rate is attained.

Third phase: This last phase of uniformity indicates dynamic balance.

How are these very characteristic results to be explained? They find a rational explanation from the physiological standpoint.

The first phase.—The pulsatory activity of every section of the midrib, which is the principal channel for the transport of sap, was in a state of arrest due to drought. The absorption of water at the lower end revived the activity at that point and gradually at others higher up, as the liquid pumped up reached them. In the course of time a large number of cellular pumps became activated, with resulting increase in the rate of ascent.

The second phase.—The propulsion of sap, as previously explained, takes place from the more to the less activated end of the organ. The lower irrigated end of the midrib being activated by turgor-distension, the flow of sap is towards the less active upper end, which becomes turgid and distended by the accession of water pumped up from below. The differential activity which determines the rate of flow thus undergoes a diminution, inducing a lowering of the rate.

The third phase.—Incipient drought at the upper end, caused by transpiration, maintains the ascent of sap at a rate which is uniform but lower than the rate at the beginning. This uniform lower rate in the third phase is attributable to the transpiration from leaves. What would happen if transpiration were totally abolished?

Experiment 27. *Transport of sap in varnished midrib.*—A leaf of *C. moschata* was taken and subjected to considerable drought. The lamina was then removed and the midrib coated with vaseline; after making a fresh cut at the lower end, it was mounted on the Potograph. The preparation was hanging down at 135 degrees from the vertical. In Experiments 25 and 26 there was transpiration from the surface of the leaf; in the present experiment, however, there was complete abolition of transpiration. In spite of this the vaselined midrib exhibited active suction, and the maximum rate, attained after 2 minutes, was 2·3 cub. mm. per minute. A typical example of the characteristic variation in the rate of ascent is given in Table VII.

TABLE VII.—RATES OF SUCTION OF THE VASELINED MIDRIB OF THE LEAF OF *CENTAUREA MOSCHATA*.

Interval after application of water	Rate of suction in cub. mm. per minute
2 minutes after	2·3
5 ,, ,,	1·8
10 ,, ,,	1·2
15 ,, ,,	0·92
20 ,, ,,	0·48
25 ,, ,,	0·30
35 ,, ,,	0·13
45 ,, ,,	0·00

Suction of water caused continuous erection of the midrib from the drooping to the vertical position.

Few results could be more decisive than the above in proving the existence of a propulsive activity independent of transpiration, and of differential activity in the maintenance of the unidirectional flow of sap. In the present case there was no loss of water by transpiration; the turgor-tension at the two ends of the organ became equalised in the course of 45 minutes, with the result of total arrest of the ascent.

Effect of Plasmolytic and of Poisonous Solutions

Experiment 28. *Effect of* KNO_3 *solution on suction and movement of leaf.*—An erect and fully turgid leaf of Cornflower was taken, and its rate of suction was found to be 2·02 cub. mm. per minute. A 5 per cent. solution of KNO_3 was substituted for the water in the Potograph. There was a rapid diminution in the rate of ascent, which in the course of 10 minutes declined to 0·69 cub. mm. per minute. The effect of the diminished rate of ascent was independently manifested by the responsive fall of the leaf from the vertical to the horizontal position of 90 degrees. After 40 minutes the rate declined still further to 0·30 cub. mm. per minute; the leaf also fell down vertically through 180 degrees from the erect position. Continued action of the solution produced practically an arrest 90 minutes after the application. That this arrest was but temporary was shown by the renewal of ascent after substitution of water for the KNO_3 solution. The rate of suction underwent a marked increase to 3·2 cub. mm. per minute after 20 minutes' application of water. The leaf which had fallen down vertically became erected to the horizontal or 90° position. Forty minutes after the substitution of fresh water the rate became the same as at the beginning—namely, 2·02 cub. mm. per minute. The results are given in detail in Table VIII.

The effects of plasmolytic and of poisonous solutions can be very definitely distinguished by their after-effects. It will be shown in the next experiment that poisonous solutions cause a permanent abolition of the ascent of sap, there being no renewal on substitution of fresh water (see also Experiment 22, p. 49). But in the case of KNO_3 solution there is a complete revival after supply of fresh water.

TABLE VIII.—RATES OF SUCTION OF TURGID LEAF OF CORNFLOWER WITH KNO_3 SOLUTION AND WITH FRESH WATER.

Time-interval after commencement	Rate of suction in cub. mm. per min.	Attitude of leaf
Normal rate	2·02	Vertically erect
10 minutes' application of KNO_3	0·69	Dropped 90°
40 ,, ,, ,,	0·30	,, 180°
90 ,, ,, ,,	0·01	—
20 minutes after fresh water .	3·20	Erection to 90°
40 ,, ,, ,, .	2·02	Vertically erect

Experiment 29. *Effect of poisonous solution in abolishing suction.*—This was carried out with the leaf of Cornflower in fully erect and turgid condition. After noting the normal rate of suction, a 5 per cent. solution of mercuric chloride was supplied and the change in the rate of suction observed at increasing intervals. The changing attitude of the leaf was also noted.

The normal rate was 2·9 cub. mm. per minute, the leaf being fully erect. The effect of the poisonous solution was shown by the immediate retardation of the rate of ascent as measured by suction, which was reduced to 1·92 cub. mm. per minute in the course of 2 minutes. The rate was continuously retarded, and after 25 minutes reduced to nearly two-fifths of the normal rate. The leaf had also by this time fallen down through 45 degrees. After 40 minutes the rate was reduced to one-fourth the normal, the leaf having fallen through 90 degrees. After an hour the rate had fallen to nearly one-eighth, the position of the leaf being now vertically downward. After 80 minutes there was a total abolition of suction. The lower end of the leaf was then washed and mounted once more in the Potograph, fresh water being substituted for the poison. The water-index now remained perfectly stationary, proving the abolition of the ascent of sap by the irreversible death-change caused by the toxic agent. The detailed results are given in Table IX.

TABLE IX.—DIMINISHING RATE OF SUCTION WITH POISONOUS SOLUTION OF $HgCl_2$ (CORNFLOWER).

Interval after application	Rate of suction in cub. mm. per minute	Angular position of the leaf from vertical
Zero	2·9	0° or vertical
2 minutes after	2·4	—
25 " "	1·2	45°
40 " "	0·7	90°
60 " "	0·4	180°
80 " "	0·0	—

Experiment 29A.—A parallel experiment was carried out with leaf of *C. moschata*, which gave very similar results. The normal rate of suction was 5·8 cub. mm. per minute. Application of poison caused a continuous diminution of the rate. After 30 minutes the rate was reduced to 3 cub. mm., and after 50 minutes to 1 cub. mm., per minute. After an hour there was permanent abolition of suction. An even shorter application of the poison may be effective in causing the death of the plant.

SUMMARY

Electric stimulation retards the rate of ascent of sap, there being a revival after a period of rest.

The rate of ascent of sap, as measured by the rate of suction, becomes immediately depressed by the action of a poisonous solution. This is independently demonstrated by the results obtained by the Tapping Record, by the Automatic Record, and by the Method of the Potograph.

The actions of plasmolytic and of poisonous solutions are clearly distinguished by their respective after-effects. Though both produce an arrest of the ascent, yet the effects on the substitution of water are very different.

After plasmolysis, the supply of water revives the ascent. But the effect of poisonous solutions is irreversible, ascent of sap being permanently abolished by the toxic agent.

VI.—EFFECT OF THE TEMPERATURE OF THE WATER OF IRRIGATION ON THE RATE OF ASCENT OF SAP

BY

J. P. Sircar, B.Sc., M.B., Ch.B.

The rate of rhythmic pulsation is slowed down by a fall of temperature, while a thermal rise enhances the frequency of pulsation. There is thus a practical arrest of the heart-beat of the frog at or about 0° C., while the frequency of its pulsation increases, as the temperature rises up to an optimum. The pulsation of the Desmodium leaflet is likewise quickened under a thermal rise, while it is arrested at a thermometric minimum.[1] Does the rate of ascent of sap exhibit an enhancement during a thermal rise, while it shows a depression or arrest during a thermal fall? The observation of such results would offer very strong proof that vital rhythmic activity is an important factor in the propulsion of sap. The following experiments were made with the object of securing a reply to this question by the observation of the movements of leaves under variation of external conditions.

The Attitude of the Leaf under Irrigation with Cold and Warm Water

Experiment 30.—The experiment was carried out at the end of February, when the average temperature of the room had risen to 24° C., and the hygrometric condition was dry. On subjecting the leaf of Cornflower to moderate drought, it drooped vertically down to 180 degrees. When

[1] *Motor Mechanism of Plants* (1928), pp. 263–267.

water at 24° C. was applied to the cut end of the leaf, it became vertically erected in the course of about 4 minutes, which was the average time for a number of similar specimens. When kept supplied with water, the vertical attitude was maintained for many hours, the loss of water by transpiration being made up by the absorption at the cut end.

Experiment 31.—In a parallel series of experiments the leaf was in a state of moderate drought and had fallen through 180 degrees. On application of water at 22° C., it began to erect itself within 30 seconds, being raised through 45 degrees in the course of 1 minute. Ice-cold water at 4° C. was next quickly substituted, with the result that the erectile movement came gradually to a stop; after this, the leaf began to droop as if supply of water had been withheld.

Cold water retards or arrests the ascent of sap.—Though water was available, yet on account of the physiological depression induced by cold, the power of suction had become more or less paralysed.

Experiment 32. *Revival of ascent by supply of warm water.*—The fact that it was the ice-cold water that caused depression of the power of suction, resulting in the fall of the leaf, is clearly demonstrated by the effect on the same leaf of warm water at 35° C.: the hitherto arrested power of suction became revived, as was manifested by the full erection of the leaf.

Temporary paralysis of the power of suction is induced when the lower end of the leaf is subjected for a comparatively short length of time to ice-cold water. Prolonged subjection to the action of cold is liable to induce a more persistent paralysis, for the leaf is then in many cases unable to erect itself after the supply of warm water.

These characteristic reactions occur very quickly when the leaf, in a state of feeble drought, is at an inclination of 90 degrees to the vertical. The following results may be taken as typical.

Experiment 33.—When the cut end of the leaf was supplied with water at 20° C., it began to rise almost immediately from its horizontal position. It rose 10 degrees above the horizon in the course of 16 seconds. Application of ice-cold water caused an arrest of the erectile movement in the course of about 10 seconds. The leaf then drooped, falling through 10 degrees in the course of 70 seconds. On substitution of warm water, the responsive fall or negative response was converted into an erectile or positive response in about 10 seconds. The rate of positive response increased continuously, and the leaf became raised from a few degrees

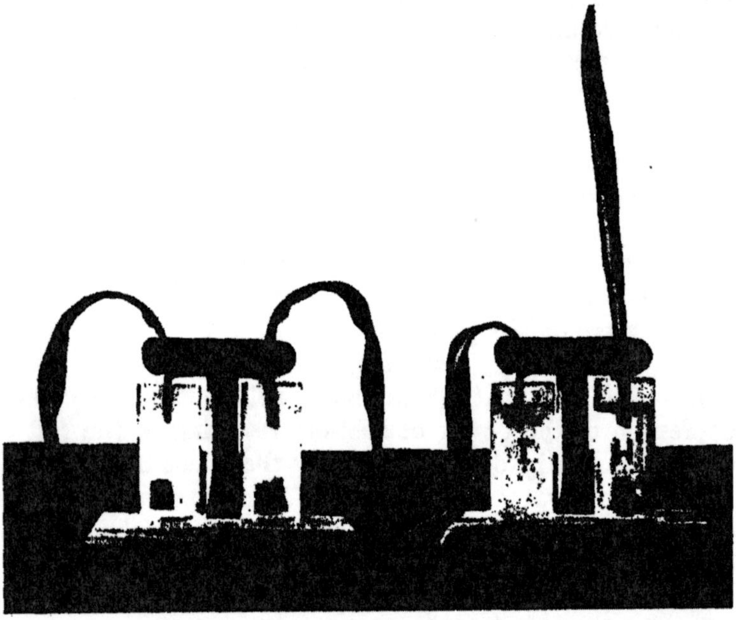

FIG. 23. Effects of supply of ice-cold water (I) and warm water (II) on similarly drooping leaves of Cornflower.
Illustration to the left shows the position before, to the right after, supply of ice-cold and of warm water.

below the horizon to the fully erect vertical position in the course of 2 minutes.

The results obtained by the independent method of Automatic Record are in every way similar (*see* Experiment 38).

Experiment 34. *Opposite effects of ice-cold and warm water.*—Two similarly drooped leaves of the Cornflower in a state of moderate drought were taken. The cut end of the leaf to the left was supplied with ice-cold water, while that to the right was irrigated with water at 35° C. The leaf to the left under the action of ice-water (I) drooped downwards, while the leaf under the action of warm water (H) erected itself vertically in a very short time (fig. 23).

TABLE X.—EFFECT OF TEMPERATURE OF THE WATER OF IRRIGATION ON A SEMI-DROOPED LEAF (CORNFLOWER).

Temperature of water	Rate of responsive
20° C.	Erectile movement initiated in 2 seconds; erection through 10 degrees in 10 seconds.
4° C.	Arrest of erectile movement followed by a fall.
35° C.	Revival of positive erectile response in 10 seconds; full erection attained in 2 minutes.

The next series of experiments was carried out by the Tapping Method, which gives record of the time-relations of the movement.

THE METHOD OF THE ELECTRO-MAGNETIC TAPPER

This records the rate of fall or of rise of a leaf which has drooped to an inclination of about 30 degrees more or less (*cf.* fig. 9). A very active leaf, which becomes quickly erected, gives no time for observing the effect of variation of temperature. In order to avoid the difficulty, a less active old leaf was employed.

Experiment 35. *Difference of suctional activity at 20° C. and 35° C.*—The record in fig. 24 is a typical example of the effect of variation of temperature on the rate of erectile movement. The successive spacings in the two records at 20° C. and 35° C. represent the time of erection for each degree of space. During the same interval of time there is an erection of 7 degrees at 20° C. and 17 at 35° C. The activity at 35° in this case is therefore 2·4 times that at 20°. In a second case, the activity at the higher temperature is found to be 3 times that at 20° C.

Experiment 36. *The relative activities at 12° C. and 20° C.*—The first record was taken when the irrigating water

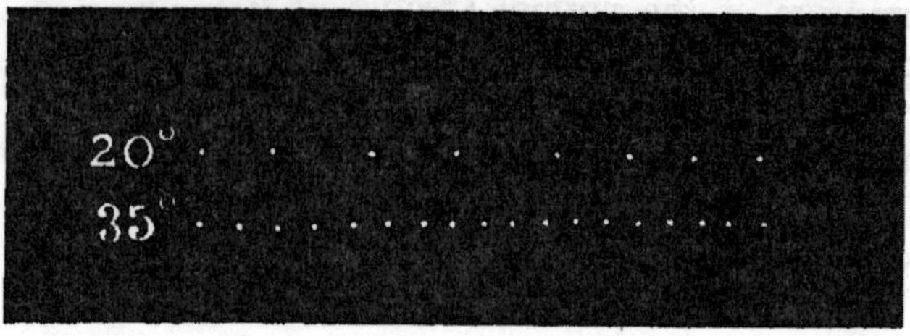

FIG. 24. Tapping records of rates of erection of leaf supplied with water at 20° C. (upper row) and at 35° C. (lower row). Successive spacings indicate times of erection per degree of space (Cornflower).

was at 12° C., and the second after substitution of water at 20° C. (fig. 25). The ratio of the two activities is as 2·2 : 1—that is to say, the activity at 12° C. is less than half that at 20° C.

The results of Experiments 34, 35, and 36 show that there is a continuous diminution of the rate of ascent

FIG. 25. Tapping records of effect of application of water at 12° C. (upper row) and at 20° C. (lower row). Rate at 12° C. is 6·5 seconds per degree; at 20° C. it is 3 seconds per degree (Cornflower). Successive spacings in this and in following figure represent time of erection for each degree of space.

under fall of temperature. If the activity at 35° C. be taken as unity, the rate is reduced to one-third at 20° C., and to one-sixth at 12° C. It would thus appear that a further lowering of temperature would cause a great depression or arrest of activity. This was actually shown when ice-cold water brought about the fall of the leaf (*cf.* Experiment 31). The critical temperature of arrest would be different in tropical and subtropical species of plants.

The depression of the rate under cold, causing a transformation of the normal erectile positive to the negative response of fall, as also the revival of activity by application of water at ordinary temperature, is demonstrated in an interesting manner by the following experiment.

Experiment 37. *Transformation of positive response into negative and re-transformation into positive.*- The leaf which

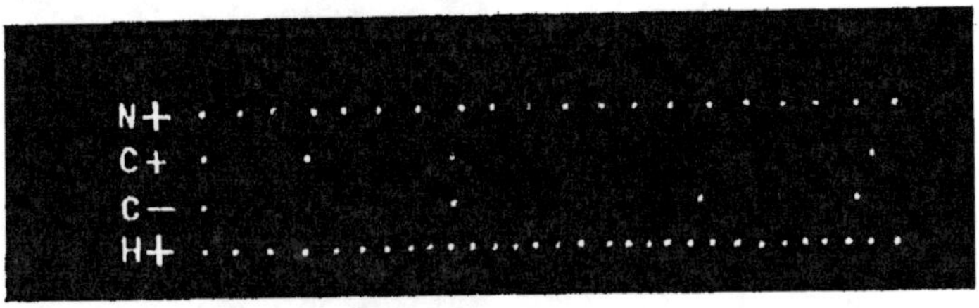

FIG. 26. Effects of application of water at 20°, at 4°, and at 30° C. (Cornflower).

N +, rate of erectile response at 20° C.;
C +, immediate slowing down of positive at 4° C.;
C −, conversion into negative after longer application of 4° C.;
H +, reconversion into positive after application of water at 30° C. (*See* text.)

was in a state of slight drought began to erect itself when supplied with water at 20° C. The normal rate of this erection, 4 seconds per degree, is shown in the uppermost record with large plus sign, N + (fig. 26). Application of ice-cold water at 4° C. caused first a slowing down of the erection, as seen in the widening of successive spacings in the second record with its smaller plus sign, C +. Continued action of cold now reversed the positive erectile response into a negative one of fall, as shown in the third record, C −. Water at 30° C. was next supplied, which transformed the negative into positive in the lowermost record, H +. The rate of erection at the higher temperature is 2 seconds per degree, which is twice as rapid as the normal rate N + at 20° C. This enhanced rate becomes still greater after a period of application. It should be remembered that, under the particular condition of the experiment, only a short time could be allowed for observation of the effect of the application of warm water.

THE METHOD OF AUTOMATIC RECORD

This method has the advantage of continuation of the record for a longer period. The leaf at an inclination of 90 degrees under feeble drought was taken: the erection of the leaf under irrigation was now easily recorded.

Experiment 38. *Effect of application of water at 22° C. and 4° C.* A leaf of Cornflower had drooped slightly under feeble drought. Application of water at the temperature of the room, 22° C., initiated an erectile movement in the course of about 10 seconds, shown in the first part of the record. Ice-cold water was then substituted, with the result that the rate of erection was slowed down, ending in the arrest of movement. Application of water at 30° C. now revived the activity of ascent, shown by the rapid erectile movement of the leaf (fig. 27).

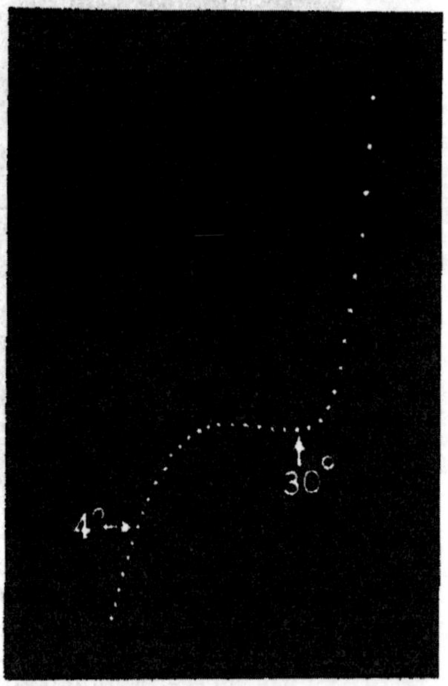

FIG. 27. Automatic records of effects of application of water at 22°, at 4°, and at 30° C. on slightly drooping leaf (Cornflower).

The erectile response retarded and arrested after application of water at 4° C. at horizontal arrow. Renewal of erectile movement after application of water at 30° C. at up-pointing arrow.

It may be said in general that the time required for the initiation of various changes—the beginning of suction on application of water, the retardation and arrest after application of cold water, and the revival of ascent is correspondingly modified by the degree of the state of drought of the specimen. If the drought has been very slight, the changes induced are exhibited within a comparatively short time.

Experiment 39. *Revival of arrested activity.*—Here the leaf had been subjected to ice-cold water and was slowly

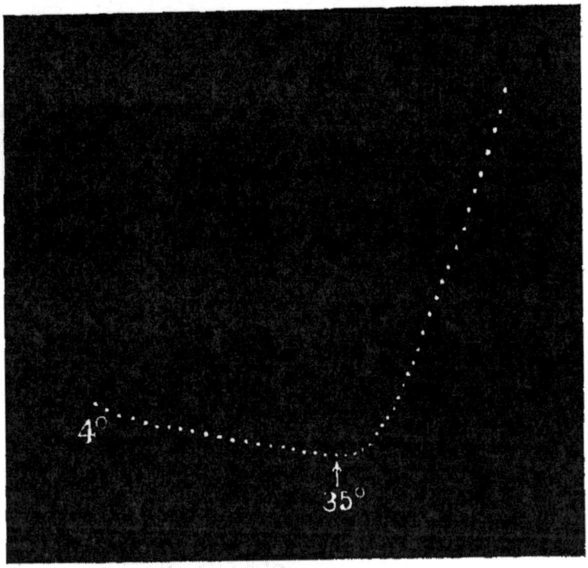

FIG. 28. Effect of ice-cold water and of water at 35° C. (Cornflower).

Owing to arrest of suction under cold at 4° C., the leaf exhibited a fall. Application of water at 35° C. transformed the drooping negative into positive erectile movement.

drooping. Application of water at 30° C. transformed the negative into positive erectile response in the course of 16 seconds (fig. 28).

Experiment 40. *Effects of application of cold, of heat, and second application of cold.*—The record was obtained with *Centaurea moschata.* The first part of the record shows drooping caused by arrest of suction at 4° C.; the second part, the revival of activity by application of water at 35° C.; and the third part of the record exhibits the stoppage of erection and subsequent fall of the leaf by the retardation and arrest of suction under the action of cold (fig. 29).

The principle of the method employed above is the detection of the variation in the rate of ascent by the concomitant change in the rate of the erectile response of

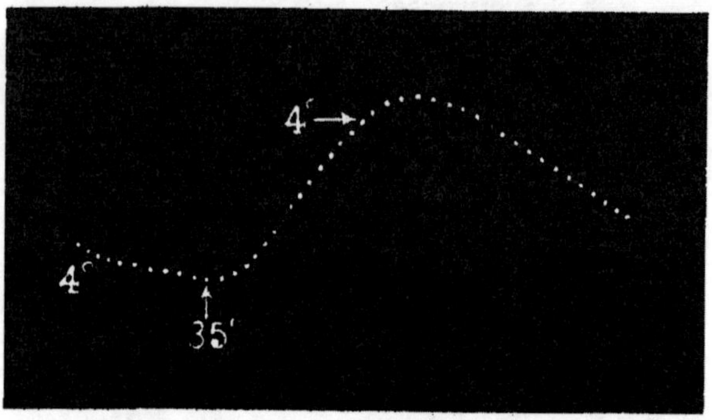

FIG. 29. Effect of successive applications of water at 4°, at 35°, and at 4° C. once more (*Centaurea moschata*).
The leaf drooping under cold became erected under heat, drooping once more under cold.

the leaf. Even more striking is the direct method of quantitative measurement of the rate of suction, indicating the rate of ascent of sap, and its variation induced by change of temperature.

The Potographic Method

The principle and working of this method has already been described (fig. 22, p. 50). An important condition in the present investigation is the maintenance at a uniform temperature of the water in the vessel after a definite thermal variation—either a fall or a rise. This is secured by encasing the Potograph in an insulating cover of thick felt, or by placing it inside a box filled with mica-dust. The bulb of a thermometer is inserted into the Potograph for reading the various temperatures, and for finding out whether a particular temperature is maintained constant for the 5 or 6 minutes during which each set of observations is made. The experiment is commenced with water at, say, 12° C. having been introduced into the Potograph. The rate of suction soon becomes uniform, as found from successive readings of the water-index at intervals of 1 minute. The temperature is then gradually increased by successive steps, water at the definite temperatures being introduced for the purpose.

Experiment 40A. *Effect of cyclic variation of temperature on the rate of suction.*—Observations were taken of the rates of suction for definite temperatures during thermal fall and also for the same temperatures during thermal rise. The results obtained in a typical case are given in Table XI.

TABLE XI.—EFFECT OF CYCLIC VARIATION OF TEMPERATURE ON THE RATE OF SUCTION (*CENTAUREA MOSCHATA*).

Temperature	Rate in cub. mm. during thermal rise	Rate in cub. mm. during thermal fall	Mean rate in cub. mm.
12° C.	1·16 ↓	↑ 1·18	1·17
15° C.	1·62	1·68	·65
20° C.	2·60	2·70	·65
24·5° C.	3·50	3·55	·52
30° C.	5·10	5·12	·11
25° C.	8·75	8·75	8·75

The most important points to be noted in the results summarised in the above table are :

(1) The rate for any particular temperature during thermal rise is approximately the same for the same temperature during thermal fall.

(2) The rate of suction undergoes a continuous fall on lowering of the temperature, while there is a continuous enhancement of the rate on rise of temperature.

Similar results were obtained in numerous other cases; the very characteristic reactions can only be regarded as the effect of physiological variation on the rate of ascent of sap.

The rates of suction at different temperatures measured by the Potograph Method are very direct and accurate.

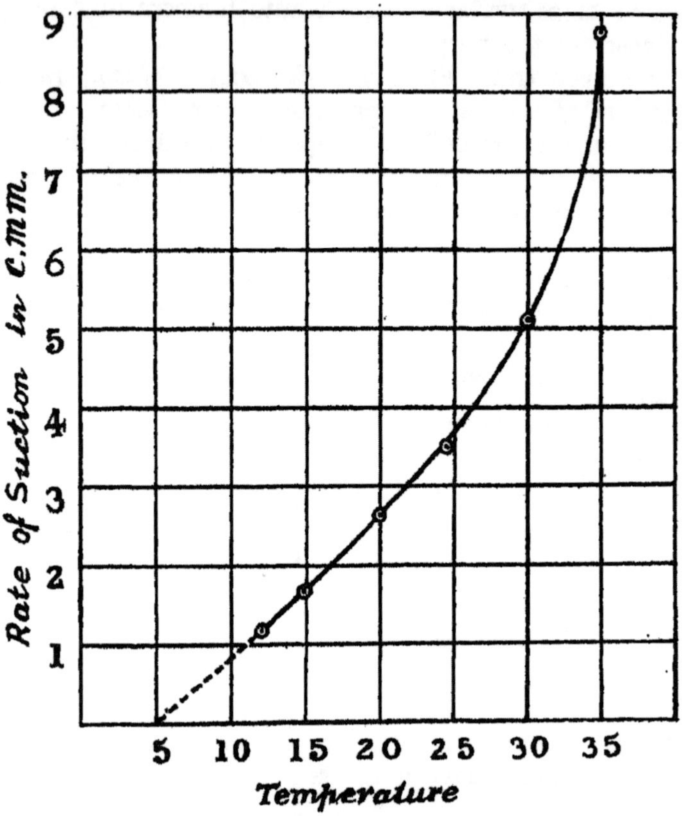

FIG. 30. Curve showing the relation between temperature and rate of suction (*Centaurea moschata*).

The curve when produced backwards cuts the abscissa at or about 5° C.

It is interesting to compare the results obtained by the Potograph with those of the Tapping Method. According to the results of the Potographic Method as given in Table XI, the rate of ascent at 20° C. is about 2·2 times greater than at 12° C. Again, the suctional activity at 35° C. is more than three times that at 20° C. The

Tapping Record gives results which are very similar. The activity at 20° C. was found to be more than twice that at 12° C., while that at 35° C. was three times greater than that at 20° C. (*cf.* Experiments 35, 36).

From the mean rate of suction at different temperatures given in Table XI a curve has been drawn (fig. 30), in which the abscissa represents the temperature and the ordinate the rate of suction. The very smooth and regular curve shows that the rate of suction increases rapidly above 30° C. When the curve is produced backwards, it cuts the abscissa at about 5° C. at a point which indicates practical arrest of suction. The various results previously given prove that the drooping of the leaf had already been initiated at 4° C. The rate of suction at that temperature must have been so reduced as not to be able to make up the loss by transpiration. This practical arrest must have occurred at a slightly higher temperature than 4° C., in fact at 5° C., as would appear from the characteristic curve given in fig. 29. On account of the prevailing high temperature of the room, 24° C., at the end of February, it was found impossible to maintain the temperature of the water in the Potograph at 5° C. But earlier in the month, when the temperature of the room was usually much lower, the water in the vessel could be maintained at 5° C., when suction was found to have been annulled.

Summary

The activity of absorption, and with it that of the propulsion of sap, has been found to be definitely modified by variation of temperature.

Investigations have been carried out by the independent methods of Tapping Record, of Automatic Record, and of the Potograph; and the various results obtained are in full agreement with each other.

The suctional activity, on which depends the responsive movement of the leaf, is increased under rise, and decreased under fall, of temperature.

A rise of the temperature of the irrigating water from

12° C. to 20° C. enhances the activity twice; the increase of activity from 20° C. to 35° C. is about threefold.

Lowering of temperature causes a great diminution of the rate of ascent. In tropical plants it becomes greatly depressed or even arrested at 4° C. Though water is present to be sucked up, yet on account of the physiological depression induced by cold, the power of suction is practically paralysed, and the leaf, instead of rising, exhibits a fall.

Supply of warm water revives the power of suction and the ascent of sap.

The results obtained by the direct measurement of the rate of suction by the Potograph show that there is a continuous increase in the rate of ascent during the thermal rise, while there is a continuous depression during the thermal fall.

The variations of the activity of suction are exhibited in a striking manner in the very smooth curve which shows the relation between the temperature and the rate of ascent.

When the curve is produced backwards, it cuts the abscissa at about 5° C. This is found to be the temperature at which the power of suction of Cornflower grown in the tropics is practically annulled, as shown by the responsive fall of the leaf.

The effects of variation of temperature upon the absorptive activity and propulsion of sap in the leaf, as described in this paper, cannot be accounted for as being produced in a purely physical manner. They are clearly the response of the protoplasm of the living cells of the organ.

VII.—RHYTHMIC PULSATION IN PLANTS

BY

SIR J. C. BOSE

THE ascent of sap has been shown to be due to vital activity of a rhythmic pulsatory character. A clear understanding of the subject is rendered possible by the study of the effects of changes of environment on pulsatory activity in general, the enhanced or depressed rates of which are visibly manifested by the waxing or waning of the associated movements. Is the rate of ascent of sap modified by any of the conditions which modify these visible pulsatory movements? Should this prove to be the case, the fact would afford weighty evidence that pulsatory activity is an essential factor in the rise of sap.

I have described elsewhere [1] investigations carried out on the automatic pulsatory movements of the leaflets of *Desmodium gyrans*, these being due to the rhythmic variation of turgidity in the cells of opposite sides of the pulsating organ. Additional experiments under new conditions are described in the next paper, and demonstration given of the recorded variation of amplitude of pulsation serving as a reliable index of changes in rhythmic activity. Experiments have been described in previous papers of this volume in which the erectile movement of various leaves has been utilised to detect interference with the ascent of sap, by drought, by variation of temperature, by plasmolytic and by poisonous solutions. A valuable independent method of detecting variation in the rate of ascent of sap is that of

[1] Bose, *Irritability of Plants* (1913), pp. 290, 341; *Motor Mechanism of Plants* (1928), chaps. xix, xx.

the observation of the modification of the movements of the Desmodium leaflet induced by such agents. The essential points of inquiry relating to this subject are :

(1) The dependence of automatic activity on energy derived from external stimulation.
(2) The relation between the ascent of sap and the pulsating activity of the leaflet.
(3) The effect of variation of internal pressure on the pulsatory movement.
(4) The effect of carbonic acid gas.
(5) Effect of irrigation by water at low and at moderately high temperature.
(6) The effect of plasmolytic withdrawal of water.
(7) The effect of poison.

The general consideration of automatic pulsation in plants suggests the important question whether the pulsatory activity of the Desmodium leaflet is an isolated phenomenon, or whether there are organs of other plants which exhibit it, though in a less pronounced degree. I have occasionally observed this in the leaflets of *Mimosa pudica*. I was, however, led to suppose that regular pulsatory movements similar to those of Desmodium might be discovered in the lateral leaflets of *Cajanus indicus*. Mr. A. Guha-Thakurta was entrusted with the investigation described in a later paper (No. IX) in this volume.

VIII.—THE RELATION OF THE PULSATORY MOVEMENT OF THE LEAFLET OF DESMODIUM TO THE ASCENT OF SAP

BY

SIR J. C. BOSE, F.R.S., AND S. C. DEB, B.Sc.

DESMODIUM is, generally speaking, an annual plant which flourishes from spring till the end of autumn, not surviving the winter months. Occasionally some specimens grow into bushes and live for three or four years. The leaves drop off in winter, and young leaflets begin to appear in spring: it is with such leaflets that these experiments have been made. The young leaflet is very small, and it was only possible by very delicate apparatus to record its rhythmic activity. The arm of the magnifying lever of an Oscillating Recorder is attached by a cocoon thread to a point about the middle of the leaflet by means of a drop of shellac varnish. As the pull exerted by the leaflet is very feeble, the writer has to be made extremely light.

On isolation from the plant, the shock of the operation was found to have arrested the activity of the leaflet, and it was only after 4 or 5 hours' rest that the rhythmic movement revived. During the process of revival the pulsations were at first irregular: an interesting phase is that in which there is a periodic alteration in the amplitude, a large pulsation being followed by a small one (fig. 31), which continues for a certain length of time till the pulsations become uniform.

The orbit of the pulsatory movement, normally a straight line, is sometimes circular or elliptical; this is due to the action of light of unequal intensity on the two flanks of the pulvinule. When the leaflet is so adjusted that its length is parallel to the direction of the light which enters through the window, the elliptical orbit is transformed into a vertical up-and-down movement of the leaflet, shown in the record as successive pulsations of equal amplitude.

The petiole bearing the leaflet is mounted watertight in the shorter open end of a narrow U-tube filled with water. The cut end of the petiole, immersed in water, is about 8 mm. below the pulsating leaflet. The longer limb of the U-tube is for the introduction of various

FIG. 31. Periodic alternation of pulsatory movement in Desmodium. A large pulsation followed by a small one.

solutions at the cut end of the petiole; a stopcock allows the water to run out of the U-tube, making way for any required solution. The effect of various solutions applied at the cut end of the petiole on the activity of the leaflet can thus be recorded.

When it is necessary to study the effect of various gases on the pulsatory activity, a small light case with mica windows is made which encloses the leaflet. The top of the chamber is also closed, except for a small circular opening for the passage of the cocoon thread which is attached to the writing lever. An inlet pipe allows a given gas to enter the chamber, the excess escaping through the outlet pipe. The gas may be blown away by the introduction of fresh air, so that an uninterrupted record can be obtained (fig. 32).

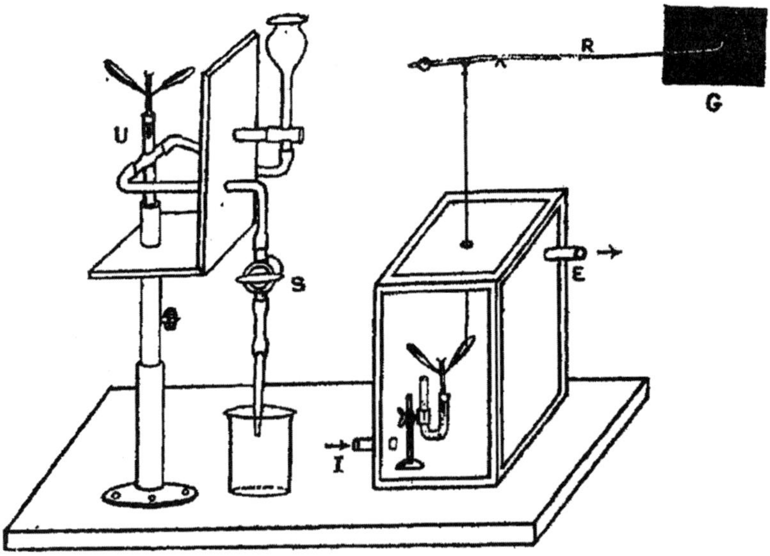

FIG. 32. U-tube U for mounting the leaflet, with stopcock S for introduction or renewal of solutions (left figure).
Mica case with inlet and outlet pipes I, E, for subjecting leaflet to different gases (right figure); R, recording lever; G, smoked-glass plate.

External Stimulation and Automatic Pulsation

The pulsatory activity of Desmodium is not spontaneous, but results from energy absorbed from external stimulation. This is proved in two different ways:

(1) By keeping the plant in the dark when depletion of its store of energy brings the pulsation to a standstill.

(2) By the revival of pulsation when the specimen is subjected to fresh stimulation.

Experiment 41. *Effect of prolonged darkness.*—The experiment was carried out in a room which was at first maintained in perfect darkness. The record was commenced in the morning, when the plant had already been subjected to the darkness of the night; the pulsations became feebler and feebler till they came almost to a stop, as seen in the first part of the record (fig. 33).

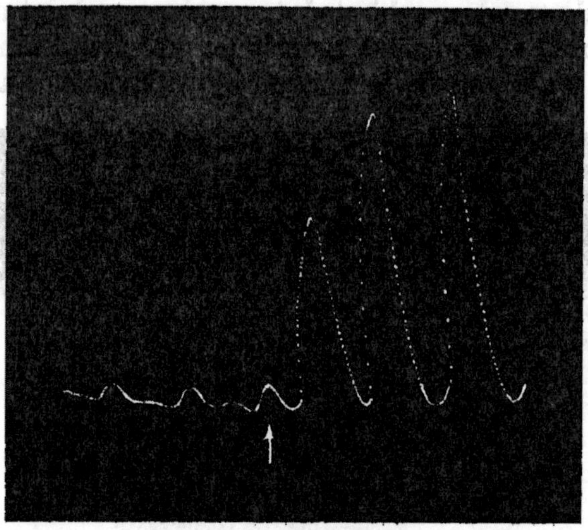

FIG. 33. Effect of stimulus of light in enhancing pulsatory activity.

The pulsation came nearly to a state of standstill in prolonged darkness. Application of feeble light at arrow revived the activity in a staircase manner.

Experiment 42. *Effect of stimulus of light.*—The shutter of the window was opened, so as to allow very feeble light to enter the room; the absorbed energy from photic stimulation was sufficient to revive the pulsations which are seen to increase in a staircase manner in the second part of the record. There can be no more decisive demonstration of the dependence of pulsatory activity on absorption of energy from external stimulation than the arrest and revival of pulsation by the withdrawal and reapplication of light.

This revival of activity can also be produced by electrical stimulation.

Effect of Drought and of Irrigation

A certain amount of internal hydrostatic pressure is necessary for the maintenance of all rhythmic pulsations. It is well known that the quiescent heart of a Snail begins to pulsate when subjected to an increased intra-cardiac pressure. The internal hydrostatic pressure of the Desmodium leaflet can be diminished by withholding water from

Fig. 34. Effect of drought and of irrigation on the pulsation of Desmodium.

First pulsation was normal; water withdrawn at down-pointing arrow resulted in a preliminary enhancement, followed by depression and arrest of pulsation. Irrigation at up-pointing arrow caused a revival.

the cut end of the petiole. The condition of drought and reduced internal pressure is removed by fresh irrigation.

Experiment 43. *Effect of withholding water and subsequent irrigation.*—A short length of the petiole of Desmodium carrying the leaflet was mounted in the U-tube so that the cut end of the petiole was in contact with water, and record of its uniform pulsation taken. Water was withdrawn by the opening of the stopcock, the petiole and leaflet being thus subjected to increasing drought. When the length of the petiole was short, the effect of the lack of water on the pulsating leaflet became evident in the course of a few minutes, the results appearing in two stages. The immediate effect due to slight drought was often an enhancement of activity, which lasted only for a short time. After this the depressing effect set in, culminating in an arrest of pulsation. Subsequent application of water at the cut end then renewed the ascent of sap along the petiole which, reaching the pulvinule, increased its internal pressure and revived the pulsation (fig. 34). A very interesting fact, sometimes observed, is that the amplitude of the revived movement may even be greater than the normal at the beginning. The amplitude, however, became normal after a certain interval of time.

The leaflet of Desmodium may thus be regarded as a visible indicator of the ascent of sap along the petiole. Retardation of the ascent is followed by an enfeeblement of the amplitude of the pulsatory movement. Conversely a diminished amplitude of movement indicates a diminished supply of water to the pulvinule through retardation of the rate of ascent; increased amplitude of movement, on the other hand, indicates an enhanced rate of ascent of sap.

The Effect of Carbonic Acid Gas on Pulsation

A supply of oxygen is necessary for the catabolic processes evolving the energy for the maintenance of all forms of pulsatory activity. This is illustrated by the following example.

A leaflet of Desmodium supplied with water maintains its pulsation for many hours; but if the leaflet is kept immersed in water, the movements are found to come to a stop in a few hours. This stoppage is evidently due to deprivation of oxygen, the amount contained in the water having become exhausted. When oxygenated water was supplied, the leaflet began to pulsate once more. The following experiment affords a direct and conclusive proof of the importance of oxygen in the maintenance of rhythmic activity.

Experiment 44. *Effect of* CO_2 *and of Oxygen.*—After taking the normal record of uniform movements, the plant-chamber was filled with CO_2. If the gas was diluted with air, a temporary enhancement of pulsatory activity was

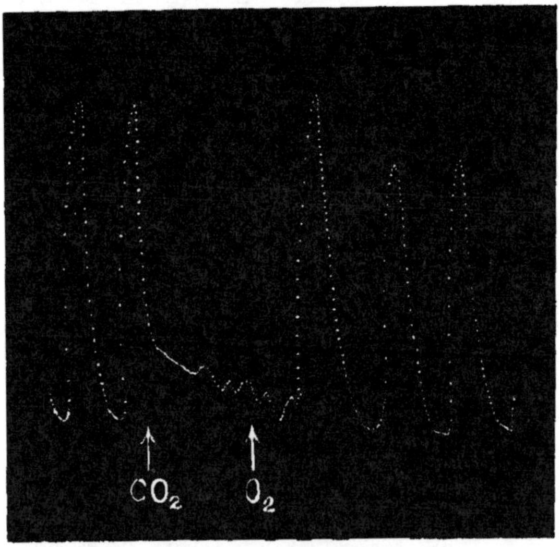

FIG. 35. Arrest of movement by asphyxiating action of CO_2. Revival on substitution of O_2.

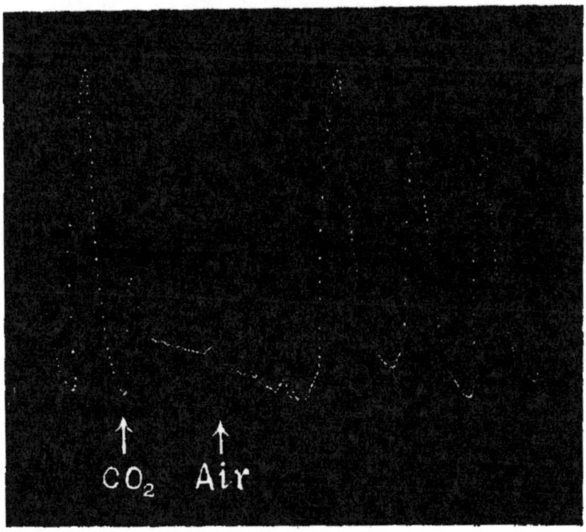

FIG. 36. Arrested movement under CO_2 revived after supply of fresh air (Desmodium).

observed; strong CO_2 caused, however, an arrest of pulsation. Oxygen was then passed through the plant-chamber, with the result of revival of the pulsatory activity (fig. 35). The experiment was continued with the same specimen, the plant-chamber being filled once more with CO_2 gas. This again caused the arrest of movement, which was revived this time by introducing fresh air into the plant-chamber (fig. 36).

Effect of Vapour of Chloroform

Experiment 45.—After taking the normal record of uniform pulsations, chloroform vapour was introduced into

Fig. 37. Effect of Chloroform vapour.
Pulsation arrested by the narcotic applied at arrow.

the plant-chamber. The effect of this strong narcotic was arrest and abolition of movement (fig. 37).

Effect of an Alternating Field of Electric Force

An investigation was undertaken to ascertain whether the automatic pulsation of the Desmodium leaflet was in any way affected by a field of electric force, generated by high-frequency alternating current. For producing high-frequency alternating current a small portable set of Tesla coil, known as the 'violet-ray apparatus,' was employed.

Two insulated plates were connected with the terminals of the secondary coil, the Desmodium leaflet being placed between the two. It is to be noted that there was no direct connection between the coil and the leaflet, the latter being placed within the space which was undergoing rapid electric variation. It is important in this connection to refer to

FIG. 38. Effect of alternating electric field applied at arrow on pulsation of Desmodium.

Note preliminary enhancement followed by arrest.

the effect of alternating electric field on the automatic activity of growth. It has been shown that the effect is modified by the effective intensity of stimulation. Under feeble stimulation by the alternating field, growth exhibits an enhancement; prolonged or strong stimulation, however, induces retardation in the automatic activity of growth.[1]

Experiment 46.—It is interesting to find that the effect of an alternating field on the automatic movement of the

[1] Bose, *Growth and Tropic Movements of Plants* (1929), p. 187.

Desmodium leaflet is very similar to that on growth. After taking normal record, the leaflet of Desmodium was subjected to the action of the alternating field. The effective stimulation depends on the intensity and duration of application. The stimulation may therefore be regarded as feeble at the beginning, and strong after long-continued action. The results are clearly demonstrated by the record in fig. 38, which shows a preliminary enhancement of pulsatory activity followed by decline and arrest.

Effect of Irrigation by Water at Low and at Moderately High Temperature

Experiment 47.—The cut end of the petiole was placed in water at the temperature of the room, 22° C., and the

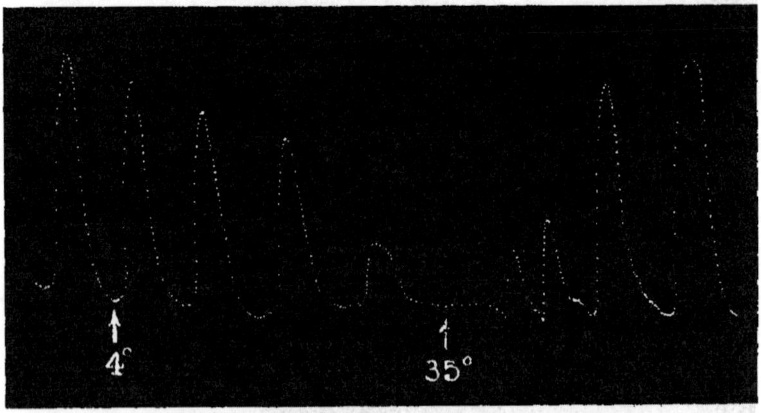

Fig. 39. Effect of irrigation with cold water at 4° C. and with warm water at 35° C., on pulsation of Desmodium. Pulsation stopped by cold; warm water caused revival of pulsation.

normal uniform record taken. Ice-cold water at 4° C. was now substituted, with the result that the pulsations became depressed almost to extinction. This must have been due to depression in the rate and arrest of the ascent of sap, as was observed by the method of erectile response (*cf.* Experiments 31, 32, 40). Warm water at 35° C. was next applied and the leaf exhibited a revived activity, the amplitude of movement being gradually increased (fig. 39). The paralysis induced by cold was thus removed by warm water. Revival of the paralysed activity was, however, difficult after prolonged application of cold. The pulsation of Desmodium comes to a stop when ice-cold water is *directly* applied on the pulvinule of the leaflet. A similar effect was observed on the *indirect* application of ice-cold water at the cut end of the petiole which bears the leaflet.

EFFECT OF PLASMOLYTIC WITHDRAWAL OF WATER

Experiment 48. *Effect of application of* KNO_3 *solution.* After taking the normal uniform record, a 5 per cent.

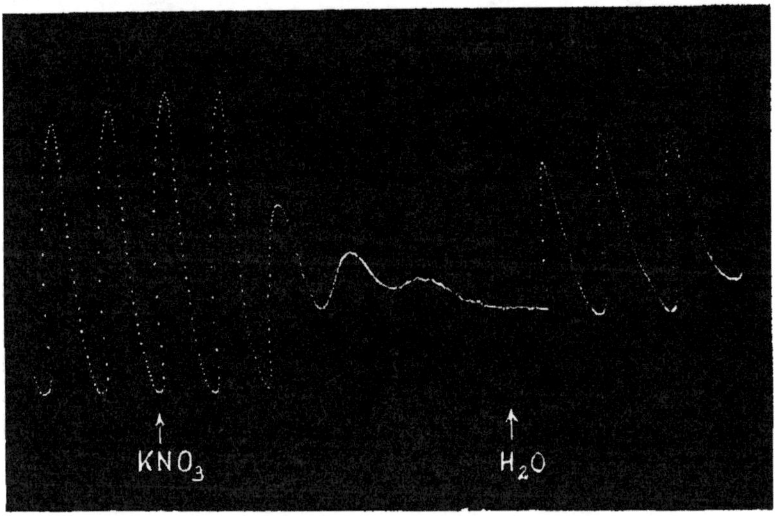

FIG. 40. Effect of plasmolytic withdrawal of water on the pulsation of Desmodium.

Arrest of pulsation by application of KNO_3 solution: revival after supply of fresh water.

KNO_3 solution was supplied to the cut end of the petiole. The pulsation now became enfeebled till it came almost to a stop. Substitution of fresh water was found to revive the activity (fig. 40).

The arrest of ascent of sap under application of KNO_3 solution, and the revival of ascent after substitution of fresh water, have already been demonstrated by the method of erectile response (*cf.* Experiments 21, 28).

EFFECT OF POISONOUS SOLUTION

Experiment 49.—After taking the normal uniform record, a 5 per cent. KCN solution was substituted for the water. This caused a diminution of the rate of ascent of

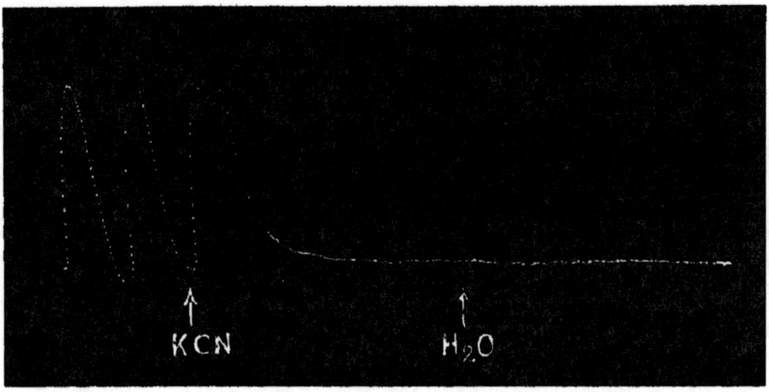

FIG. 41 Effect of poisonous solution on pulsation of Desmodium.

Pulsation arrested by KCN solution and not revived by fresh application of water at H_2O.

sap, as observed in the continuous diminution of the amplitude of pulsation culminating in final arrest. It is to be remembered in this connection that a poisonous liquid can travel through the tissue which has been killed by it and thus rendered passive. But so long as the part above remains alive, it continues to exhibit suction, which can only be abolished after every section of the petiole has been killed by the poison.

The fact of complete abolition of the power of suction in the petiole of Desmodium is proved by the effect of replacement of poison by fresh water. In the case of the KNO_3 solution there was a revival of pulsatory activity.

But after the application of the KCN solution, fresh water did not bring any revival, the rhythmic activity of the leaflet being permanently abolished by the irreversible change of death (fig. 41).

A poisonous solution thus induces an immediate retardation of the rate of ascent of sap, culminating in a permanent arrest. This has been independently demonstrated by the abolition of the erectile response of the leaf (*cf.* Experiment 22), and also by the permanent abolition of the power of suction as shown by the Potograph (*cf.* Experiment 29A).

Summary

The rhythmic activity of the lateral leaflet of *Desmodium gyrans* affords a very striking example of automatic pulsatory movement in plants. These pulsations, strictly speaking, are not automatic, but dependent on energy absorbed from external stimulation.

A diminished internal hydrostatic pressure, caused by withdrawal of supply of water, brings about a depression or arrest of the pulsation. Ascent of sap becomes renewed after irrigation and the pulsatory activity then becomes revived.

A supply of oxygen is necessary for the maintenance of pulsatory activity.

The effect of carbonic acid gas is shown in the arrest of the pulsatory movement of the leaflet. Substitution of pure oxygen or fresh air revives the rhythmic activity.

The effect of strong vapour of chloroform is an arrest and abolition of the pulsating activity.

A high-tension alternating field of electric force induces a preliminary enhancement of the automatic pulsation, followed by decline and arrest.

The effect of an alternating electric field of force on the automatic pulsation of Desmodium and on the automatic activity of growth is in every way similar.

Ice-cold water has been found to stop the ascent of sap; it also arrests the pulsatory activity of the leaflet. Substitution of warm water renews the ascent of sap and revives the pulsation of the leaflet.

Plasmolytic withdrawal of water by KNO_3 solution causes a depression and arrest of the pulsation. This arrest is temporary, for substitution of pure water revives the pulsation.

The effect of a poisonous solution is, however, different. It induces an immediate diminution in the rate of ascent which culminates in a permanent abolition. As a result of this the pulsation of the Desmodium leaflet, under the action of poison, becomes permanently arrested, no revival being possible on the substitution of pure water.

It has been shown that the pulsatory movement is affected by various agents such as cold or warm water, plasmolytic and poisonous solutions. These agents may act directly upon the pulvinule of the leaflet or indirectly when applied at the cut end of the petiole. The results here described may be attributed to interference with the supply of water, in view of the results obtained by the method of erectile response, as described in previous papers (Nos. V and VI).

IX.—DISCOVERY OF RHYTHMIC PULSATORY MOVEMENT IN THE LEAFLET OF *CAJANUS INDICUS*

BY

ASHUTOSH GUHA-THAKURTA, C.S.AG.C.

AUTOMATIC pulsation is exhibited in a very conspicuous manner by the lateral leaflet of *Desmodium gyrans*. Is this instance unique, or is it possible that similar activity may

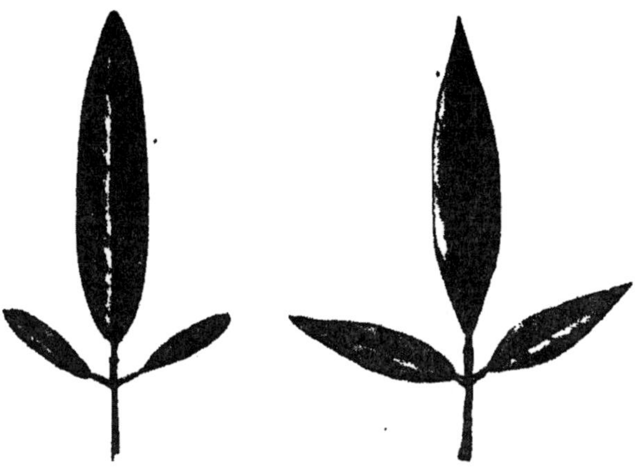

FIG. 42. Leaf of Desmodium (left illustration); leaf of Cajanus (right illustration, reduced to half natural size).

be discovered in the leaves of other plants? Sir J. C. Bose kindly suggested an investigation on the subject with the leaf of *Cajanus indicus*.

There is a considerable resemblance between Desmodium and Cajanus. They belong to the same natural order Leguminosae, and to the same sub-order Papilionaceae.

The flowers are very much alike. In both plants the leaf is compound, a terminal large leaflet with two small lateral ones. The Desmodium leaf is shown in its natural size in the left illustration, the leaf of Cajanus reduced to half the normal size is shown in the illustration to the right (fig. 42). The difference in the two leaves is in the size of the lateral leaflets: in Desmodium the leaflets are small and light, while in Cajanus they are large and heavy.

It is the small lateral leaflets of Desmodium that exhibit pronounced automatic pulsatory movement, the period of each of which varies from 3 to 5 minutes according to the season and the prevailing temperature, the movement being quickened by a thermal rise and slowed by a thermal fall. The terminal large leaflet of Desmodium does not, however, exhibit any definite pulsatory activity. In what way, now, does the organisation of the terminal leaflet differ physiologically from that of the lateral?

Response of Terminal Leaflet of Desmodium to Stimulation

The terminal leaflet of a moderately young leaf of a potted Desmodium plant was attached to the writer of an Oscillating Recorder, the successive dots being at intervals of 1 minute. The plant and apparatus were placed in an open veranda which was uniformly illuminated. The leaflet began to exhibit rapid nyctitropic movement in the afternoon which interfered with the normal record of response to external stimulation. The best time for the experiment was, therefore, in the forenoon, when the terminal leaflet remained in the normal outspread position. A glass cover placed over the plant protected it from the disturbing action of air-currents. Stimulation was effected by an induction coil actuated by a 4-volt dry battery. One of the electrodes from the secondary coil was attached at the middle of the leaflet by kaolin-paste, while the other was connected with the petiole. The intensity of the induction shock was rendered sub-maximal by the gradual approach of the secondary to the primary coil. A metronome was employed for maintaining a uniform duration of the shock for 1 second in successive experiments.

Experiment 50.—The terminal leaflet, subjected to successive electrical stimulations of uniform intensity, exhibited responsive fall from which it recovered after a

FIG. 43. Uniform responses of terminal leaflet of Desmodium to electric stimulations applied at the thick dots.

In this and in the following records time-interval between successive dots is 1 minute.

period of rest. The up-curve represents the responsive fall, and the down-curve the recovery. The maximum fall occurred after 5 minutes, and the leaflet exhibited erectile recovery in the further course of 13 minutes (fig. 43).

EFFECT OF LIGHT AND DARKNESS

Experiment 51.—The excitability of the leaflet remains uniform under uniform light. In order to find out whether a change of illumination induces a variation of excitability the experiment was carried out as follows:

After taking the record of uniform responses under light, the glass case, within which the plant was enclosed, was covered with a black cloth. The subsequent record showed that the irritability had been depressed by change from light to darkness (fig. 44).

FIG. 44. Effect of light and darkness on moto-excitability of terminal leaflet of Desmodium.

The first two responses exhibit normal excitability under light. The next three responses show the depressing effect of darkness.

FIG. 45. Effect of CO_2 on moto-excitability of terminal leaflet of Desmodium.

The horizontal line below represents the duration of application of CO_2.

Note preliminary enhancement of excitability gradually restored on substitution of fresh air.

EFFECT OF CO_2 ON MOTO-EXCITABILITY

Experiment 52.—The leaf was enclosed in a bell-jar with a small opening at the top. A thin cocoon thread attached to the leaf passed through this opening and was tied to the recording lever.

Carbonic acid gas could be passed into the bell-jar by means of an india-rubber tube and afterwards removed by blowing in fresh air. In the record (fig. 45) the first two responses are normal. CO_2 was then introduced into the plant-chamber; the thick dot represents the moment of application of stimulus. The increased amplitude of contractile response shows that the preliminary effect was an enhancement of moto-excitability; this was followed by a great depression, as evidenced by the extreme feebleness of the next response. The original excitability was, however, gradually restored, as shown by the two responses after the introduction of fresh air.

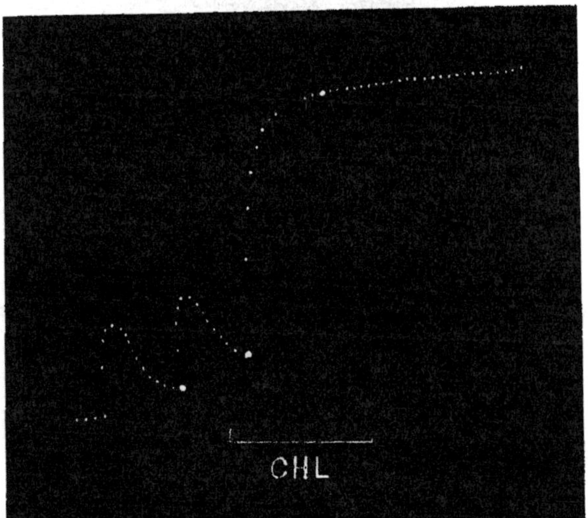

FIG. 46. Effect of Chloroform on moto-excitability of terminal leaflet of Desmodium.

Preliminary enhancement followed by abolition of excitability.

Effect of Chloroform on Moto-excitability

Experiment 53.—The effect of chloroform is shown in the record (fig. 46). There was a preliminary enhancement of excitability, which became permanently abolished by the continued action of the strong narcotic.

The above results show that the reaction of the terminal leaflet of Desmodium is similar to that of the leaf of Mimosa, the response being evoked only under external stimulation. A single stimulation gives rise to a single response; carbonic acid gas causes a preliminary enhancement of excitability, followed by depression in both Mimosa and Desmodium; finally, the very strong narcotic chloroform causes in both a preliminary enhancement of excitability, followed by a permanent abolition of the power of contractile response.

Thus, in a leaf of Desmodium two types of movement are exhibited, one by the terminal and the other by the lateral leaflets. The essential difference is that the lateral leaflets move automatically, whereas the terminal leaflet does not, but only moves when directly stimulated.

Response of the Terminal Leaflet of *Cajanus indicus*

A parallel series of experiments was carried out with the terminal leaflet of Cajanus. The petiole bearing the leaf was isolated from the parent plant and suitably mounted in a U-tube; the water in the longer limb of the U-tube exerted a moderate hydrostatic pressure on the petiole. To simplify matters the lateral leaflets were cut off, and the terminal leaflet was attached to the recording lever. The specimen was placed within a moist chamber and the normal excitability was found to be restored after a rest of 1 hour.

Response to Electrical Stimulation

Experiment 54.—The response and recovery were found to be completed in the course of 20 minutes. The excitability, under proper conditions, remains constant for many hours. The uniformity of the responses is seen in the series of records given in fig. 47.

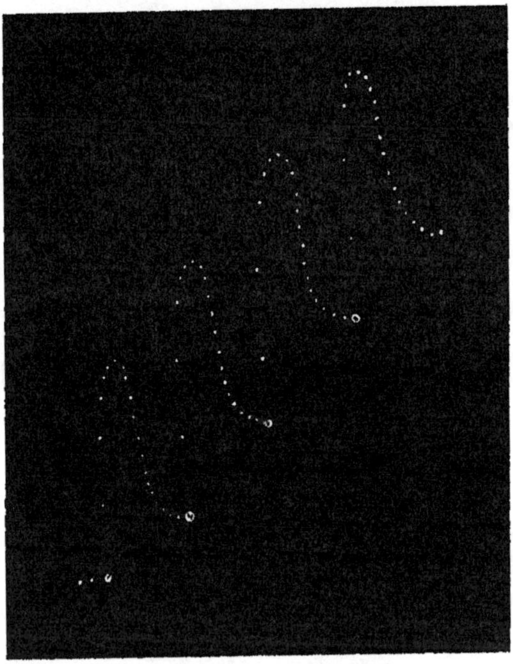

FIG. 47. Uniform response of terminal leaflet of Cajanus under electrical stimulation.

EFFECT OF CO_2 ON MOTO-EXCITABILITY

Experiment 55.—After taking two uniform responses under sub-maximal stimulation, CO_2 was slowly passed into the plant-chamber. This caused, as in the terminal leaflet of Desmodium, a preliminary enhancement followed by a great depression of excitability (fig. 48).

EFFECT OF CHLOROFORM

Experiment 56.—Two normal responses were first recorded, after which chloroform vapour was introduced into the plant-chamber. After a preliminary enhancement,

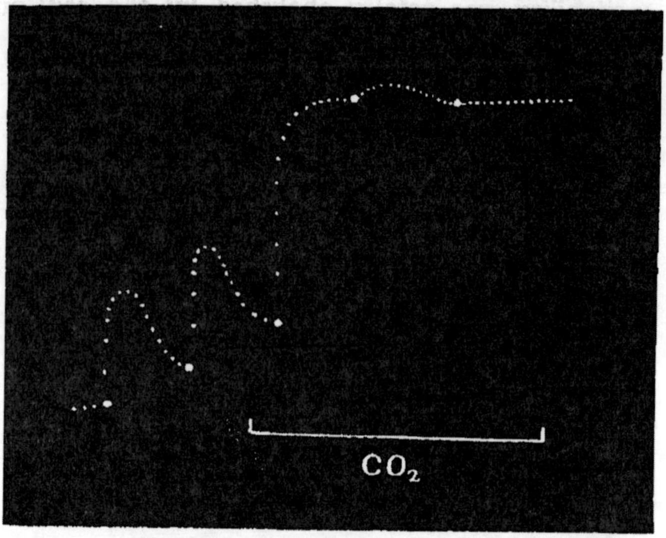

FIG. 48. Effect of CO_2 on moto-excitability of terminal leaflet of Cajanus.

Preliminary enhancement followed by depression of excitability.

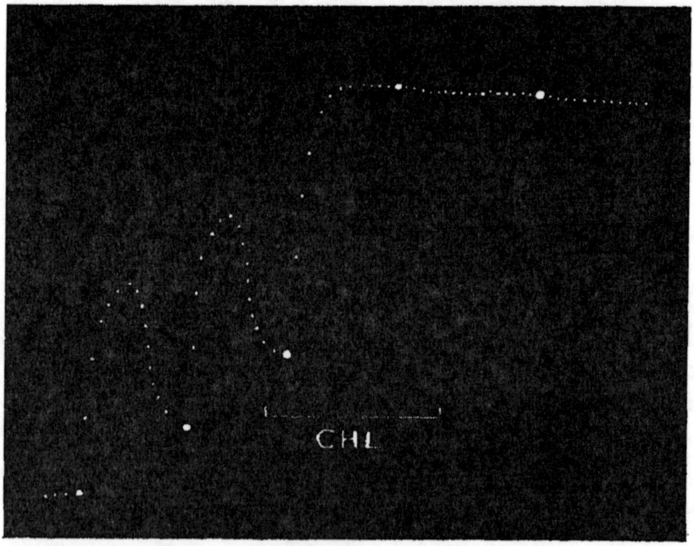

FIG. 49. Effect of Chloroform on terminal leaflet of Cajanus. Preliminary enhancement followed by abolition of excitability (*cf.* fig. 45).

the excitability was totally abolished by the continued action of chloroform. Introduction of fresh air did not produce any revival of response (fig. 49).

Having proved that the response of the terminal leaflet of Cajanus is similar to that of Desmodium, the attempt was next made to discover the nature of the response of a lateral leaflet.

Careful observation of successive positions of a lateral leaflet appeared to show that it was not stationary, but was executing a periodic up-and-down movement. One could not, however, be perfectly sure whether it was really automatic or caused by the disturbing action of currents of air. Arrangements were therefore made to prevent mechanical disturbance and secure automatic record of the movement of the leaflet.

The Cajanus plant, which was in a state of vigorous growth in the garden of the Institute, was covered with a large glass case. Direct sunlight was cut off by a white screen so that uniform diffused light acted on the plant. A leaf, which was neither too young nor too old (being the fourth one counted from the top), was selected for the experiment. The main petiole was held by a universal clamp, so that the movement of the lateral leaflet, if any, would take place in a vertical plane.

Record of Movement of Lateral Leaflet

Experiment 57.—The Oscillating Recorder was so adjusted that the smoked glass plate moved through 10 cm. in the course of 2 hours. The successive dots in the record indicate intervals of 1 minute. The record was taken in December between 1.30 P.M. and 3 P.M., the average temperature inside the plant-chamber being 23° C.

The record revealed a remarkable automatic movement of the lateral leaflet of Cajanus which is essentially similar to that of the lateral leaflet of Desmodium (fig. 50). The lateral leaflet of Desmodium is, as previously pointed out, very small, whereas that of Cajanus is large and heavy. This may perhaps explain the relative slowness of the pulsation in Cajanus. In Desmodium the period of a single pulsatory movement in the cold season is about 5 minutes, whereas in Cajanus it is six times longer, being 30 minutes for each movement. The uniformity of the successive movements of the lateral leaflet of Cajanus is very remarkable. This uniformity is maintained so long as the external conditions, such as light and temperature,

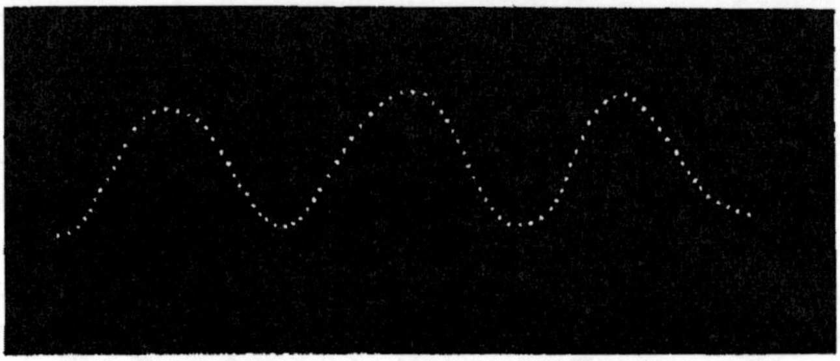

FIG. 50. Uniform automatic movements of lateral leaflet of Cajanus.

are approximately constant. Towards the evening, light wanes and temperature undergoes a rapid decline, with the result that the movements become enfeebled and irregular.

SUMMARY

In Desmodium two types of moto-excitability are exhibited: the lateral leaflets exhibit automatic pulsatory movements, whilst the terminal leaflet moves only in response to external stimulation.

The following are the characteristics of the response of the terminal leaflet of Desmodium:

(1) Under uniform electrical stimulation the responses are uniform. The response and recovery are completed in the course of about 18 minutes.

(2) Light is necessary for the maintenance of normal excitability of the leaflet. Depression of excitability occurs on keeping the leaflet in darkness.

(3) Carbonic acid gas, after causing a preliminary enhancement of excitability, brings about depression. Excitability is, however, restored on reintroduction of fresh air into the plant-chamber.

(4) Long-continued action of chloroform vapour causes abolition of excitability.

Very similar are the characteristics of the terminal leaflet of Cajanus. Thus uniform responses occur under uniform electrical stimulation, the period of response and recovery being about 20 minutes. Carbonic acid gas induces a preliminary enhancement of excitability followed by great depression. Finally, vapour of chloroform causes abolition of excitability.

The lateral leaflet of Cajanus, like that of Desmodium, exhibits automatic pulsatory movements. The period of each of these is about six times longer than that of Desmodium.

The discovery of the automatic pulsatory movement of the lateral leaflets of Cajanus shows that the phenomenon is not peculiar to the lateral leaflets of Desmodium but is of wider occurrence.

X.—THE CONDUCTION OF IMPULSE IN PLANTS

BY

SIR J. C. BOSE

FOR the purpose of establishing the universality of the phenomenon of transmission of excitatory impulse in plants, the scholars of my Institute have been invited to repeat under new conditions the experiments previously performed, and also to carry out a long series of investigations on every species of 'sensitive' plant that could be obtained.

A brief reference to the researches on the conduction of impulse, already described in my 'Nervous Mechanism of Plants' (1926), will be helpful in understanding the scope of the additional investigations on the subject detailed in several of the following papers. The conclusion that the transmission of impulse in the plant is, in fact, the travelling of protoplasmic excitation, is justified by the following facts.

I have been able to prove that, like the nervous impulse in the animal, an impulse is initiated in the plant by the discriminative polar action of an electric current. With a feeble current excitation is produced in both only at the kathodic point on the make of the current; this first type of reaction is simply expressed by the symbolic formula Km. With a moderate current, excitation is initiated not only on the *make* of the current at the kathode, but also on the *break* of the current at the anode, this second type of reaction being represented by the formula KmAb. It is to be inferred that the excitatory impulse, thus initiated and afterwards conducted to a distance, is purely physiological and not merely physical.

In regard to the special tissue for the conduction of impulse in plants, it has been shown (*ibid.*, p. 176) that separate conducting strands in the vascular bundles extend through the sub-petioles of the leaf of Mimosa to the central end, the pulvinus, where they become coalesced into a continuous ring. When a stimulus is applied to the first sub-petiole situated to the left, the impulse generated reaches the pulvinus and causes the fall of the leaf. After the fall, an outward impulse appears to be initiated from the central end; and, proceeding in a centrifugal direction, it reaches the second sub-petiole, causing the successive closure of its leaflets in an outward direction. Stronger intensity of stimulation causes closure of the leaflets in the third and fourth sub-petioles. These and other facts indicate that it is easier for the impulse initiated at the first sub-petiole to make the double journey to the second sub-petiole *via* the central end at the pulvinus, than to travel from one vascular bundle to the other across the petiole. This partial block can, however, be overcome by a sufficiently strong stimulus, so that the impulse can get across from one bundle to the other, even after the removal of the pulvinus. The phenomenon is then analogous to the axon reflex in the animal.

Experiments which I carried out on the action of stronger exciting currents on plants showed that with a moderately strong current, stimulation is caused not only at kathode-make and anode-break, but also at anode-make, the formula of this Type III being KmAmAb. A still stronger current gives rise to excitation at the make and break of kathode, and at the make and break of anode. The excitation of the Type IV is thus KmKbAmAb. The existence of Types III and IV has been denied and attempts made to explain them by the assumption of the production of secondary poles by the strong current employed. Arrangements were, however, made which eliminated the possibility of the production of secondary poles. Nevertheless, the definite reactions of Types III and IV were still found to occur.[1]

If the Laws of Polar Excitation are the same in the conducting tissue of the plant and in the nervous tissue of the animal, then these two new types of reactions III and IV should also be found to exist in the animal nerve. A new investigation on this subject is given in one of the papers of this series (p. 120).

Crucial evidence in proof of the excitatory character of the transmitted impulse is afforded by the arrest caused by various physiological blocks. Additional experiments will be described in confirmation of the previous results.

It has been a debated question whether chloroform is effective in retarding or arresting the impulse in plants. If so, in what way is the effect modified by the strength and duration of application? For the comparative study of the effect of chloroform in plant and in animal, the characteristic effect of chloroform in producing a partial or complete block was first studied, as described in one of the papers of this series, with the animal nerve, in order to observe, under parallel conditions, if a similar block was produced in the conducting path of the plant (p. 135).

[1] Bose, *Motor Mechanism of Plants* (1928), pp. 114, 115.

XI.—THE TRANSMISSION OF IMPULSE IN BOTH ASCENDING AND DESCENDING DIRECTIONS

BY

B. K. Dutt, B.Sc.

By the kind suggestion and advice of Sir J. C. Bose I have been encouraged to carry out investigations on the character of the transmitted impulse in plants. The methods employed are those described in his 'Nervous Mechanism of Plants' (1926).

Stimulation of the stem of *Mimosa pudica* initiates an impulse which is transmitted to a distance, causing the fall of the sensitive leaves. It has been suggested that this is not due to the transmission of an excitatory impulse, but is brought about by a movement of the sap. This theory is based on the supposition that a hypothetical stimulant, excreted as the result of irritation of the wood by a deep wound, is carried to the leaves by the transpiration-current.

The fact that, without causing any deep wound, various modes of stimulation of the stem give rise to impulses which travel *simultaneously* both upwards and downwards, the one in the direction of the transpiration-current, the other against it, is sufficient to disprove the theory that the transpiration-current is concerned in the transmission of impulse. The nature of the impulse can be fully ascertained from the various characteristics associated with it; more specially by the initiation of the impulse by purely physiological means, and also by the arrest of its transmission by a physiological block. Certain important aspects of the subject will be taken up in the following order:

I. The transmission of impulse simultaneously upwards and downwards.

II. Downward propagation of impulse on stimulation of the apex of the organ.

III. The effect of unilateral stimulation.

Transmission of Impulse simultaneously Upwards and Downwards

The effect of stimulation, applied at the middle of the stem, was observed, the successive fall of the leaves above and below indicating the passage of the impulse. Stimulation was produced by electrical or thermal shocks.

Electrical stimulation.—Two electrodes of an induction coil were applied on the epidermis, secure electric connection being made by means of kaolin-paste moistened with normal saline. The intensity of the stimulation could be increased from the minimal to the maximal by moving the secondary coil nearer to the primary.

Thermal stimulation.—A loop of thin platinum wire surrounded the stem of the plant. The passage of an electric current through the thin wire heated it short of incandescence. This caused stimulation by thermal shock.

Experiment 58. *Simultaneous transmission in two directions.*—Application of tetanising induction electric shocks at the middle of the stem of *Mimosa pudica* gave rise simultaneously to two impulses, one of which travelled upwards in the same direction as the transpiration-current; whereas the other travelled downwards *against* the direction of the transpiration-current. It is to be noted that in this method:

(1) There was no wound to cause excretion of the hypothetical stimulant in the wood.

(2) The two impulses generated by the stimulus travelled simultaneously in opposite directions. The impulse downwards, as already pointed out, travels against the direction of the transpiration-current.

The velocities of the two impulses were determined :

Transmission in the upward direction

Distance of transmission .	50 mm.
Time required . . .	15 seconds
Velocity in upward direction	3·3 mm. per second

Transmission in the downward direction

Distance of transmission .	40 mm.
Time required . . .	37 seconds
Velocity in downward direction	1·1 mm. per second

Experiment 59. *Simultaneous transmission under thermal shock.*—The velocity of impulse is in general increased with stronger intensity of stimulation, and under the stronger stimulus of thermal shock the velocity is found to be relatively greater. In this case also the speed is quicker upwards than downwards.

The propagation of the impulse downwards completely invalidates the theory that the transpiration-current effects its transmission.

The velocity of transmission in the stem of Mimosa is found to vary from 2 to 10 mm. in a second. It is appropriately modified by the physiological vigour of the specimen, by the season, by the temperature and by the intensity of the stimulus.

Though the fact of simultaneous transmission cannot be explained by the theory of the transpiration-current, yet it finds a complete and rational explanation by regarding the phenomenon as one of conduction of excitatory impulse analogous to that in the nerve of the animal. For impulse is conducted in the animal nerve in both directions at the same time, as has just been shown to be the case in the conducting tissue of the plant. In the plant there appears to be a greater facility of conduction in one direction than in the other, since the velocity in the upward direction is greater than that in the opposite direction.

Downward Propagation of Impulse on Stimulation of the Apex

The apex of the organ was stimulated in order to demonstrate the transmission of impulse against the direction of the transpiration-current. Various species of sensitive plants were employed for this purpose, such as *Mimosa pudica* and two varieties of *Mimosa Spegazzinii*, one of which is a shrub, the other tree-like in habit here described as the Tree Mimosa. The main petiole of the Tree Mimosa bears about seven pairs of opposite sub-petioles, which are sensitive and respond to stimulation by a fall. The leaflets on the sub-petioles also respond to stimulation by a movement of closure.

Experiment 60. *Downward transmission in the petiole of Tree Mimosa.*—Strong stimulation of the apex of the petiole was produced by the momentary contact of a heated wire. This caused successive fall of the pairs of sub-petioles, beginning with the first and ending with the seventh pair. The following is a summary of typical results obtained with two different specimens:

Transmission in Downward Direction (Petiole of Tree Mimosa)

Specimen 1.	Distance of transmission	135 mm.
	Time required . .	25 seconds
	Velocity . . .	5·4 mm. per second
Specimen 2.	Distance of transmission	155 mm.
	Time required . .	27 seconds
	Velocity . . .	5·7 mm. per second

Experiment 61. *Downward transmission in the stem.*—This was carried out with the stem of the shrubby form of *M. Spegazzinii.* The sensitive leaves are borne alternately on the stem. Application of strong thermal shock at the apex caused successive fall of leaves from above downwards. The following are typical results:

Transmission in Downward Direction (Stem of M. Spegazzinii)

Specimen 1.	Distance of transmission	125 mm.
	Time required . .	15 seconds
	Velocity . . .	8·3 mm. per second
Specimen 2.	Distance of transmission	75 mm.
	Time required . .	14 seconds
	Velocity . . .	5·3 mm. per second

EFFECT OF UNILATERAL STIMULATION

In straight and untwisted stems of *Mimosa pudica* and of *M. Spegazzinii* the leaves are in alternate series. The leaves of the odd series on the right are vertically over each other; this is also the case with the leaves of the opposite even series on the left.

There are two opposite main bundles in the stem of *Mimosa pudica* and of *M. Spegazzinii* which give off lateral branches to the odd and even series of the leaves. The two main strands of the conducting tissue converge and nearly meet with each other at the apex, so that the resistance offered for the passage of the impulse across the top is comparatively slight; in addition to the two main or principal strands in the stem there are intermediate ones, the conducting tissues of which are not continuous with but contiguous to those of the two principal strands. A considerable resistance is thus offered to the spread of excitation in a lateral direction. Hence the effect of unilateral stimulation of moderate intensity is conducted lengthwise along the line of least resistance, and not laterally across the stem. It is only when the intensity of the stimulus is greatly increased that the excitatory impulse becomes transmitted across the stem. What has been said of the stem holds good in regard to the conduction of impulse in the main petiole of Tree Mimosa.

In the 'Nervous Mechanism of Plants' (1926) the effect of unilateral application of mechanical stimulation of increasing intensity on the stem of *Mimosa pudica* has been described. Electrical stimulation was also employed for the same purpose. At Sir J. C. Bose's suggestion I employed the electrical method, which possesses important advantages. The experiments were carried out not only with the stems of *Mimosa pudica* and *M. Spegazzinii*, but also with the main petiole of the Tree Mimosa.

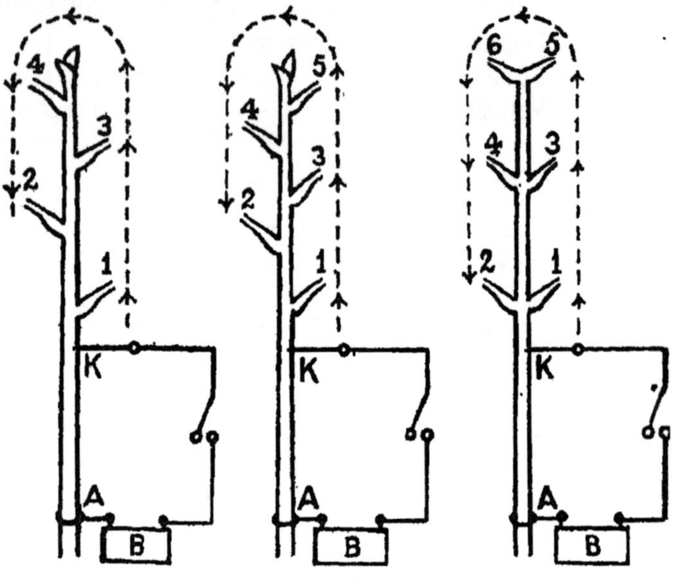

FIG. 51. Unilateral electric stimulation applied (*a*) to the stem of *Mimosa pudica*; (*b*) to the stem of the shrubby form of *M. Spegazzinii*; and (*c*) to the petiole of Tree Mimosa.

B, battery. Current started by pressing key. A, anode; K, kathode.

Note ascending impulse on the right side, crossing over to the left, and its subsequent transformation as a descending impulse (*cf.* Experiments 62, 63, 64).

The electrical stimulation was effected by the discriminative polar action of a constant current.[1] The impulse was initiated at the kathode on the make of the current; the induced excitation varies with the strength of the current and the excitability of the specimen. In winter, when the specimen is in a less excitable condition, the effective stimulation is ensured by raising the E.M.F. acting on the circuit.

[1] Bose, *Nervous Mechanism of Plants* (1926), p. 27.

Suitable electrical connections were made with the stem. The lower connection, the anode, was made by means of a moist band of cloth. For the kathode a fine platinum pin was pricked in slightly so as to make a better connection the main conducting strand. After a suitable period of rest, the excitatory electric current was started by pressing a key.

The excitation at the kathode due to a moderate E.M.F. caused a transmission of impulse through a short distance, the leaf immediately above undergoing a fall. A stronger impulse was generated by the employment of a higher voltage. It is to be remembered that the resistance of the circuit was often very high, and the employment of a strong E.M.F. was necessary to ensure the initiation of sufficiently strong impulse.

Experiment 62. *Unilateral stimulation of stem of Mimosa pudica.*—In this particular case the experiment was carried out in the winter month of December, when the conducting power was feeble. The strong impulse at the kathode-make, produced by an E.M.F. of 40 volts, ascended on the right side and caused the leaves to fall in sequence from below upwards. After reaching the apex, the impulse crossed over to the left side, the direction of propagation becoming reversed from ascending on the right side to descending on the left (fig. 51). The results are given below, K being the point of stimulation :

Intervening distance		Successive time-intervals
From K to 1	. . 10 mm.	35 seconds
,, 1 to 3	. . 27 mm.	43 ,,
,, 3 to 4	(*via* apex) 20 mm.	17 ,,
,, 4 to 2	. . 28 mm.	85 ,,

Average velocity of ascending impulse 0·63 mm. per second
,, ,, ,, descending ,, 0·33 ,, ,, ,,

Thus under effectual unilateral stimulation the impulse ascending on one side along the main conducting strand, after reaching the apex, crosses over to the opposite side and is then converted into a descending impulse. This cannot be explained by an hypothesis such as the transmission of a hydro-mechanical impulse, or that of the transport of a stimulant by the sap. The propagation along definite conducting strands proves, on the other hand, that it is a transmission of protoplasmic excitation. The above experiment on transmission of impulse in *Mimosa pudica* under unilateral electric stimulation is a repetition under different conditions of what has already been described in the 'Nervous Mechanism of Plants.' The following new experiments were carried out with other species of plants.

Experiment 63. *Unilateral stimulation of stem of M. Spegazzinii.*—The application of an E.M.F. of 50 volts initiated an impulse which ascended on the right side, causing successive fall of the leaves; after reaching the apex it crossed over to the left and was then propagated in a downward direction. The results are as follows:

Intervening distance		Successive time-intervals
From K to 1 . . .	5 mm.	20 seconds
,, 1 to 3 . . .	40 mm.	38 ,,
,, 3 to 5 . . .	40 mm.	42 ,,
,, 5 to 4 . . .	30 mm.	35 ,,
,, 4 to 2 . . .	45 mm.	45 ,,

Average velocity of ascending impulse 0·9 mm. per second
,, ,, ,, descending ,, 0·8 ,, ,, ,,

Experiment 64. *Unilateral stimulation of the petiole of Tree Mimosa.*—This was carried out in spring when the conducting capacity was fairly good. There were six opposite pairs of sub-petioles. Unilateral electric stimulation caused initiation and transmission of impulse, giving rise to successive fall of the sub-petioles. The highest or No. 5 sub-petiole on the stimulated side fell 41 seconds after stimulation. The impulse then crossed over and reached the opposite sub-petiole after 14 seconds, after which it travelled in a downward direction, causing the fall of the even series of sub-petioles, 6, 4, and 2. The following is a tabular statement of results:

Intervening distance		Successive time-intervals
From K to 1 . . .	20 mm.	15 seconds
,, 1 to 3 . . .	20 mm.	5 ,,
,, 3 to 5 . . .	20 mm.	7 ,,
,, 5 to 6 . . .	0 mm.	14 ,,
,, 6 to 4 . . .	20 mm.	12 ,,
,, 4 to 2 . . .	20 mm.	11 ,,
Average velocity of ascending impulse .		3·3 mm. per second
,, ,, ,, descending ,, .		1·2 ,, ,, ,,

SUMMARY

Electrical or thermal stimulation applied at the middle of the stem of *Mimosa pudica* gives rise to two impulses which travel simultaneously in opposite directions upwards and downwards. The impulse downwards is against the direction of the normal transpiration-current.

Unilateral stimulation gives rise first to an ascending impulse which travels along the main conducting strand on one side. The impulse then crosses over at the apex to the other conducting strand on the opposite side, and continues as a descending impulse. It is not one particular organ, but any organ, such as the stem or the petiole of diverse species of plants, that exhibits this characteristic feature.

The facts just described cannot be explained by any supposition of hydro-mechanical transmission of impulse, or of any participation of the sap in the transpiration-current. The transmission along definite conducting strands, both upwards and downwards, proves, on the other hand, that the process is a travelling of protoplasmic excitation.

XII.—EXCITATION OF PLANTS BY POLAR ACTION OF AN ELECTRIC CURRENT

BY

B. K. DUTT, B.Sc.

ONE of the crucial tests in proof of the excitatory character of the transmitted impulse in plants is its initiation by the specific polar action of a constant current. The characteristic results obtained with different organs of numerous sensitive plants have been previously given in detail.[1] My instructions were to carry out additional experiments on the subject with *Mimosa pudica*, grown under various climatic conditions, as well as with the Tree Mimosa.

Mimosa pudica is grown at the Institute either in the open ground or as potted plants. It grows in moderately vigorous condition in Calcutta during the summer and autumn months from the middle of March to the end of October. It is in a moribund condition during the winter months.

The plant, however, grows very vigorously in a more favourable habitat outside Bengal, at Hazaribagh in Behar, at a distance of more than 250 miles from Calcutta. It grows there in the forest-land as a weed, its range being extended to adjoining lands, where the cultivators find it very difficult to extirpate.

Since the plants were practically dead in Calcutta during the winter, it was impossible to continue any further experiments with them. A journey was therefore undertaken to Hazaribagh, where, even in December, the plants

[1] Bose, *Irritability of Plants* (1913); *Nervous Mechanism of Plants* (1926); *Motor Mechanism of Plants* (1928).

were in a moderately good condition; the leaves were unusually large and the petioles even longer than 60 mm. It was thought desirable to carry out experiments with plants growing naturally in the forest. The prevailing wind, however, caused much mechanical disturbance; this difficulty was overcome by erecting a small circular tent of thin muslin to shelter the plants and the observer. There was good ventilation inside the tent, and the plant was uniformly illuminated by light diffused through the thin cloth. The prevailing temperature was about 15° C.

Electromotive Force for Polar Excitation

The Potentiometer.—In highly sensitive specimens such as the petiole of Mimosa in summer, the minimally effective E.M.F. for each type of reaction is comparatively low. It is therefore necessary to adjust carefully the derived E.M.F. by means of a Potentiometer slide. The current can be sent in one direction or the other by means of a Commutator or reversing key (fig. 52).

The Dry-cell Battery.—In the less sensitive specimens of *Mimosa pudica* and in the sub-petiole of the Tree Mimosa, a higher E.M.F. of 20 to 50 volts is required to initiate excitation, because the electrical resistance of the circuit is very great. The minimally effective E.M.F. is higher in the colder month of January then in spring or summer. For obtaining the necessary voltage it is extremely difficult to make use of a battery of portable storage cells, which, moreover, require frequent re-charging. The moderate current for polar excitation of plants is required for only a relatively short time. A dry cell is therefore well suited for the purpose, its internal resistance being negligible compared with the resistance of the external circuit. The compact dry battery used for wireless apparatus is found very suitable for the present purpose. It requires no special attention, and can be used for more than a month for the completion of a large number of experiments.

There are two plugs, one of which is fixed to one terminal of the battery. The second plug is movable and can be inserted at different plug-holes of the battery. In the first row of the battery terminals the E.M.F. is increased to 10·5 volts by seven steps. There are other rows in which the E.M.F. is raised by 10 volts at each step. The E.M.F. can thus be increased, say, from 5 to 60 volts. The two plug terminals are led to a Commutator or reversing key, so that the current can be sent through the plant in one

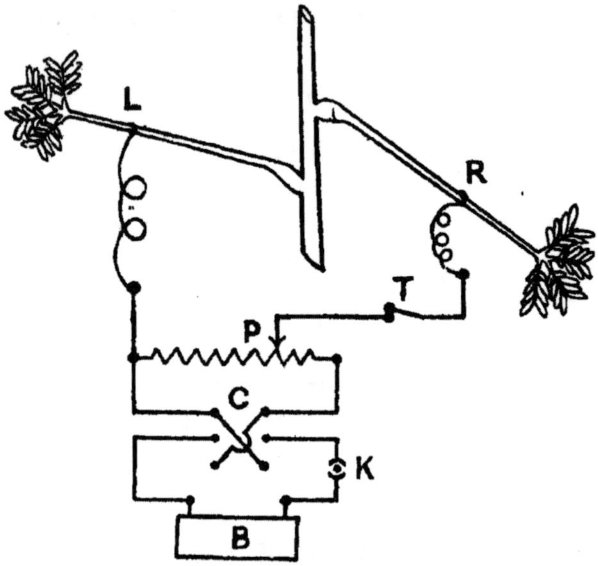

FIG. 52. The Potentiometer.

B, the battery; K, the plug-key; C, the Commutator or reversing key; P, the Potentiometer slide; T, the tapping key by which the current can be made or broken; R, the electrical connection with the right petiole; L, that with the left petiole. Illustration shows that kathode-make initiated at R an excitatory impulse which caused fall of the leaf on the right.

direction or the other. The electrical connections are made by means of two electrodes, E, E′, of flexible silver-tinsel, which are attached to the plant by kaolin-paste moistened with dilute glycerine, which prevents drying of the electrolytic contact (fig. 53).

The effects of increasing intensity of electric current in the initiation and subsequent transmission of excitatory impulse will now be described. The Bipolar Method is preferably employed for this purpose. It has the advantage that not only is the effect of kathode and anode observed at the same time, but the results are capable of further

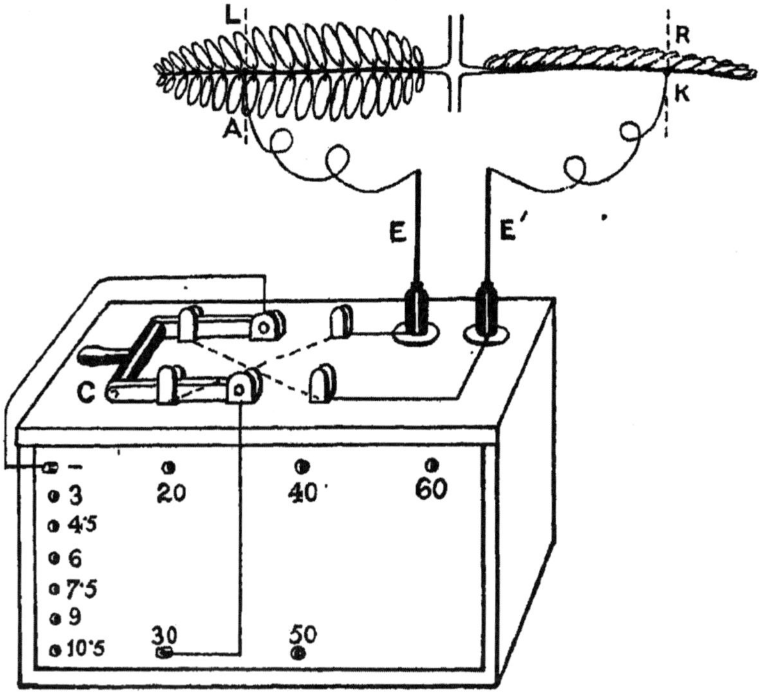

FIG. 53. The dry-cell battery with terminal connections for increasing the E.M.F. from 3 to 60 volts.

C, the Commutator or reversing key; E, E′, the two electrodes with flexible silver-tinsel to make electrical connections with the right or the left sub-petiole of Tree Mimosa.

The illustration shows that excitation was initiated and transmitted on kathode-make K at the right side R, and that there was no excitation on anode-make A at the left side L.

verification by reversal effects brought about by changing the direction of the electric current by the Commutator or reversing key.

POLAR EXCITATION OF PRIMARY PETIOLE OF MIMOSA

The irritability of two primary petioles, the second and third counted from the apex of the stem, was found to be approximately similar. Two electrical connections were made, one at the point R on the right petiole, the other at the point L on the left petiole. The intervening distance between R and the responding pulvinus to the right, and that between L and P′ to the left, were kept the same. The distance in different cases was from 20 to 60 mm.

Experiment 65. *Effect of feeble current.*—This was carried out in Calcutta in summer. An E.M.F. of 2 volts was found sufficient for minimally effective stimulation. When R was made the kathode, the excitatory impulse was conducted to the pulvinus through the intervening distance of 35 mm. after an interval of 3 seconds (see fig. 52). Ignoring the short latent period of the pulvinus, the velocity of transmission was 11·7 mm. per second. The make of the anode at L did not initiate any excitatory impulse and there was no responsive fall of the left leaf. After allowing sufficient time for the recovery of the leaf to the right, the current was reversed, L being made the kathode and R the anode. The excitatory impulse was now initiated not at the new anodic point R, but at the kathodic point L. This impulse reached the left pulvinus and caused the fall of the left leaf.

Under feeble current, excitation thus occurs at kathode-make, the formula being Km, designated as Type I.

Experiment 66. *Effect of moderate current.*—Continuing the experiment with the same specimen, the electric current was increased by raising the E.M.F. to 4 volts. Excitation was initiated and transmitted as in the last case on the kathode-make at the left. After sufficient time had been allowed for recovery, the current was broken. Excitation now occurred on the anode-break at the left and not on the kathode-break at the right. Similar results were obtained when the current was reversed, L being made the kathode and R the anode.

With moderate current, excitation occurs on kathode-make and on anode-break, the formula being KmAb, representative of Type II. The effect of further increase of E.M.F. will be described later in this paper.

In order to show that these characteristic reactions are

universal and not dependent on the particular climatic conditions prevailing in Bengal, experiments were carried out with plants growing at Hazaribagh. The character of the soil and the climatic conditions were there widely different. An interesting adjustment of the Mimosa plant to the prevailing conditions was found in the fact that, though the moto-excitability of the pulvinus practically disappeared in Calcutta at a temperature slightly below 20° C., the excitability still persisted at Hazaribagh, where the temperature in December was as low as 15° C.

Experiment 67. *Effect of feeble current on Hazaribagh Mimosa.*—An E.M.F. of 4 volts was employed, the two electrodes being applied at R on the right petiole and at L on the left. The intervening distance of transmission to the pulvinus was 20 mm.; when R was made the kathode, the right leaf exhibited a fall after 5 seconds; the velocity of conduction was thus 4 mm. per second approximately. There was no excitation of the left petiole on anode-make. The current was then reversed, L being made the kathode. The excitatory impulse was transmitted to the pulvinus after an interval of 6 seconds, the velocity of conduction being 3·3 mm. per second, which is not very different from that in the right petiole. The formula of excitation with feeble current is Km.

The E.M.F. was next increased to 6 volts. Excitatory impulse was now found to be initiated and transmitted on the kathode-make as also on the anode-break. The excitation formula with moderate current is therefore KmAb.

Polar Excitation in Sub-petiole of Mimosa

Experiments on the polar excitation were carried out with the sub-petiole of *Mimosa pudica* in repetition of those already described ('Irritability of Plants,' 1913). New investigations were also undertaken with the sub-petiole of the Tree Mimosa, the passage of the excitatory impulse being exhibited by the successive closure of the leaflets.

Experiment 68. *Effect of feeble current on sub-petiole of* Mimosa pudica.—The electrical connections were made with a pair of opposite sub-petioles, as in the diagram (fig. 54), R with the right and L with the left sub-petiole. Employing an E.M.F. of 4 volts, when R was made the kathode, all the leaflets of the right sub-petiole underwent closure in sequence. There was no excitation of the left sub-petiole on anode-make. When the current was reversed, excitation was induced in the left sub-petiole and not in the right.

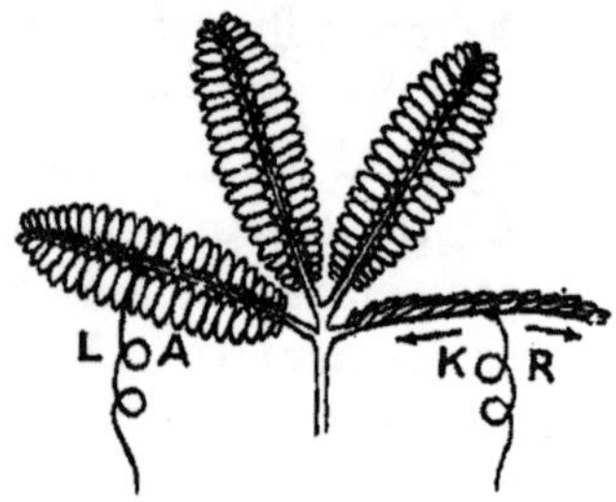

FIG. 54. Excitatory impulse in sub-petiole of *Mimosa pudica* under feeble current.

Impulse initiated on kathode-make K at the right R and not on anode-make A at the left L.

Experiment 69. *Effect of moderate current.*—The E.M.F. was next increased to 6 volts. The excitatory impulse was initiated not only at kathode-make, but also at anode-break.

In the sub-petiole of *Mimosa pudica* the excitation formula under feeble current is Km; under moderate current it is KmAb.

The next experiments were carried out with the sub-petiole of Tree Mimosa. The excitability of the specimen was considerably less than that of *Mimosa pudica*; the electrical resistance of the circuit was, moreover, very great. Hence a relatively high E.M.F. was found necessary to initiate excitation.

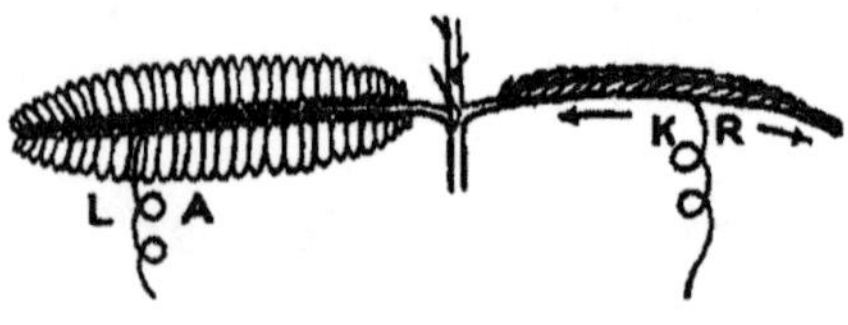

FIG. 55. Excitatory impulse in sub-petiole of Tree Mimosa under feeble current.

Reaction similar to that in fig. 54.

Experiment 70. *Effects of feeble and moderate currents on sub-petiole of Tree Mimosa.* The electrical connections were made as in the diagram (fig. 55). Employing an E.M.F. of 40 volts, kathode-make at the right point R initiated excitation which, being transmitted, caused successive closure of all the leaflets. No excitation was induced on anode-make at L on the left. Reversal of the direction of the current initiated excitation at the new kathode at L and not at the anode R.

The current was next increased by raising the E.M.F. to 60 volts. The excitatory impulse was now initiated not only on the kathode-make, but also on the anode-break.

The excitatory reactions under feeble and moderate currents are thus the same in the two species of plants and in their different organs, such as the petiole or the sub-petiole. For feeble current the excitation formula in all cases is Km; for moderate current it is KmAb.

Effect of Stronger Current

By increasing the intensity of the polar current, other types of excitatory reaction made their appearance as described in the following experiments.

Experiment 71. *Polar action of strong current in the petiole of Mimosa.*—In an experiment, previously described, with the petiole of *Mimosa pudica* by the Bipolar Method (*cf.* Experiment 67), a moderate intensity of current, produced by 6 volts, was found to initiate excitation on kathode-make and on anode-break. The experiment was continued with the same specimen, with the difference that the current was increased by raising the E.M.F. from 6 to 8 volts. On starting the current, an excitatory impulse was initiated not only at kathode-make on the right of the petiole, but also at anode-make on the left. The initiated impulse in both cases caused, after the short time required for transmission through the intervening distance, the fall of the corresponding leaves. Break of the current initiated the impulse at the anode. Excitation was thus produced on kathode-make, on anode-make, and on anode-break. The excitatory formula is therefore KmAmAb, representative of Type III.

Similar results were obtained with the sub-petiole of the Tree Mimosa, the only difference being that a stronger current had to be employed.

The following additional results fully confirm those already described in the 'Irritability of Plants' (1913).

Experiment 72. *Polar action of strong current on the sub-petiole of Tree Mimosa.*—On application of an E.M.F.

of 55 volts to two points on opposite sub-petioles, excitation was found to occur on make at both the kathode and the anode. The break of the current caused excitation at the anode. The excitation formula is therefore KmAmAb.

A still stronger current induced excitatory transmission on kathode-make, on kathode-break, on anode-make, and on anode-break, the formula being therefore KmKbAmAb of Type IV.

In the animal nerve, types of responses corresponding to the above have not been suspected. Investigations on this subject are described in the next paper.

SUMMARY

In the conducting nerve of the animal, excitatory impulse is initiated by the specific polar action of an electric current and is transmitted to a distance.

Exactly similar results are obtained in plants, impulse being initiated by the characteristic polar action of an electric current and conducted to a distance. This was exhibited by all sensitive plants, as well as by their different organs.

These specific reactions demonstrate the physiological nature of the transmitted impulse in plants.

The following are the complete Laws of Polar Excitation in Plants:

(I.) WITH FEEBLE CURRENT, EXCITATION TAKES PLACE ONLY AT KATHODE-MAKE.

(II.) WITH A CURRENT OF MODERATE INTENSITY, EXCITATION OCCURS AT KATHODE-MAKE AND AT ANODE-BREAK.

(III.) WITH A STRONG CURRENT, EXCITATION TAKES PLACE AT KATHODE-MAKE, AT ANODE-MAKE, AND AT ANODE-BREAK.

(IV.) WITH A STILL STRONGER CURRENT, EXCITATION OCCURS AT KATHODE-MAKE, AT KATHODE-BREAK, AT ANODE-MAKE, AND AT ANODE-BREAK.

XIII.—POLAR EXCITATION OF ANIMAL NERVE WITH AN ELECTRIC CURRENT OF INCREASING INTENSITY

BY

SIR J. C. BOSE AND N. N. DAS, M.B., M.SC.

IT has been shown that an excitatory impulse is initiated in plants under the polar action of an electric current, and is transmitted to a distance. With a feeble current, excitation occurs only at the kathode-make; with a current of moderate intensity, excitation takes place not only at kathode-make, but also at anode-break. These two, Types I and II, are exactly like those well known in the case of animal nerve. But under stronger intensity of current two others, Types III and IV, have been shown to be exhibited by various organs of sensitive plants. The question arises whether these two types are peculiar to plants and do not occur in the conducting nerve of the animal, or whether the reactions classified under Types III and IV can also be detected in the animal nerve.

POLAR ACTION OF ELECTRIC CURRENT OF INCREASING INTENSITY

In answer to the question that has just been raised, the following investigations were carried out with nerves of the toad, *Bufo Melanosticus*. The method employed was Bipolar, so that the reactions of kathode and anode could be simultaneously observed on the left and the right sciatic nerves of the animal. The experimental arrangement and electrical connections are diagrammatically represented in fig. 56. The spinal cord in the spinal column S was completely destroyed by pithing with a long blunt needle, so that there was no nerve-continuity between the left and right sciatic nerves N, N′, though there was electrical

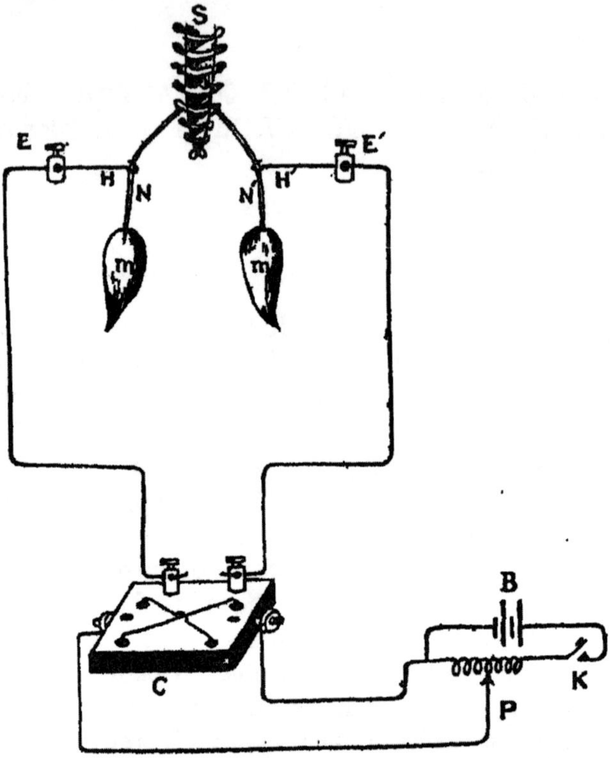

FIG. 56. Diagrammatic representation of polar excitation of animal nerve by an electric current.

N and N′ are left and right sciatic nerves supported by platinum hooks in connection with electrodes E and E′. B, battery; the applied E.M.F. adjusted by Potentiometer slide P. C, Commutator or reversing key.

continuity between the two. The sciatic nerves w exposed by dissecting away the adjacent muscles, and were supported on two platinum hooks H H′, which were connected with the two electrodes E and E′, for the electric current to enter the nerve at N and leave by N′ or *vice versa*. The object of holding the nerves in a raised position is to remove the possibility of production of secondary poles by contact with neighbouring muscles. The two electrodes are in circuit with a battery B, the derived electromotive force being capable of fine adjustment by the Potentiometer

FIG. 57. Photograph of actual experimental arrangements on polar action of current on the nerves of a Toad (*see* text).

slide P. The applied E.M.F. is measured by a milli-voltmeter not shown in the figure. The direction of the current may be changed by the Commutator C, so that the characteristic effects can be verified by making one of the two nerve-connections either kathode or anode.

The next figure (fig. 57) is from a photograph of the actual arrangement of the experiment. The unnecessary muscles have been removed by dissection, and the two legs of the animal are held above the knee-joints by two clamps, C, C, which are supported on a highly insulated rod of ebonite. The preparation is put into a moist chamber which can be closed by two glass slides, one in front and the other behind. The excitation initiated and conducted in the one or the other nerve, by the polar action of the current, was easily detected by the responding twitch of the terminal muscle to the right or to the left.

The experiment was carried out in the winter months of December and January. Any abnormal irritability of the nerve or muscle, due to dissection, was found to disappear after one hour's rest. By maintaining the moist chamber in a properly humid condition, the very definite reactions described below could be repeatedly observed for at least 2 or 3 hours.

The Characteristic Excitatory Reactions

The following series of experiments were carried out with the nerve of an identical toad, when the applied E.M.F. was gradually increased. The results were further verified by reversal effects due to the change in the direction of the current.

Experiment 73. *Effect of feeble current.*—As the electrical resistance of the circuit was constant, the intensity of the current was gradually increased by increasing the derived E.M.F., which was exactly adjusted by the Potentiometer. The E.M.F. employed was first 0·15 volt. The contact on the left nerve was at first made the kathode, and the make of the current initiated an excitatory impulse in the left nerve, causing a twitch of the left leg, while the right nerve and muscle remained unexcited. On reversing the current, the excitatory impulse was initiated at the kathode point on the right nerve and the right leg twitched in response. The excitatory formula is thus Km.

Experiment 74. *Effect of moderate current.*—The E.M.F.

was now increased to 0·37 volt. The excitatory effects were observed by sending the current first from the right to the left nerve and afterwards from the left to the right. The excitation was found to be initiated not only at kathode-make, but also at anode-break. The excitation formula is therefore KmAb.

Experiment 75. *Effect of moderately strong current.*—The E.M.F. was next increased to 0·75 volt. Excitation was now produced not only at kathode-make and anode-break, but also at anode-make. The excitation formula of this Type III is therefore KmAmAb.

Experiment 76. *Effect of a strong current.*—When the E.M.F. was raised to 3 volts, excitation occurred at kathode-make, kathode-break, anode-make, and anode-break. The excitation formula of this Type IV is thus KmKbAmAb.

It may be said in general that the minimally effective E.M.F. for each type of reaction depends on the irritability of the animal. In a highly excitable specimen, the range of E.M.F. required for exhibition of different types of reaction was from 0·15 to 3 volts. In a less excitable specimen the range was from 0·4 to 6 volts. The typical results are given in the two following tables, in which an inclined cross indicates the initiation of the excitatory impulse.

TABLE XII.—POLAR EXCITATION OF NERVE OF TOAD UNDER INCREASING E.M.F.

(Highly excitable specimen.)

E.M.F.		Km	Kb	Am	Ab	Excitation formula
Feeble current	0·15 volt	×	—	—	—	Km
Moderate ,,	0·37 ,,	×	—	—	×	KmAb
Moderately strong current	0·75 ,,	×	—	×	×	KmAmAb
Strong current	3·0 volts	×	×	×	×	KmKbAmAb

TABLE XIII.—POLAR EXCITATION OF NERVE OF TOAD UNDER INCREASING E.M.F.

(Less excitable specimen.)

E.M.F.		Km	Kb	Am	Ab	Excitation formula
Feeble current	0·4 volt	×	—	—	—	Km
Moderate „	2·0 volts	×	—	—	×	KmAb
Moderately strong current	2·5 „	×	—	×	×	KmAmAb
Strong current	6·0 „	×	×	×	×	KmKbAmAb

The above sequence is normal; occasional variation may sometimes occur, resulting from fatigue or depressed condition of the specimen.

The attention of animal physiologists is drawn to the two definite types of reaction, III and IV, which were exhibited by numerous specimens. The experimental conditions in all of them were such that the possibility of the production of secondary poles was completely excluded. In each case the effects of feeble and moderate currents were perfectly normal, coming under Pfluger's Law of Polar Excitation. But when the intensity of the current was increased, with the same nerve and with identical connections, the excitatory reactions became gradually transformed in regular sequence, first into Type III and then into Type IV.

The complete Laws of Polar Excitation in the animal nerve under increasing intensity of current are then as follows:

I. WITH FEEBLE CURRENT, EXCITATION TAKES PLACE ONLY AT KATHODE-MAKE.

II. WITH A CURRENT OF MODERATE INTENSITY, EXCITATION OCCURS AT KATHODE-MAKE AND AT ANODE-BREAK.

III. WITH A MODERATELY STRONG CURRENT, EXCITATION TAKES PLACE AT KATHODE-MAKE, AT ANODE-MAKE, AND AT ANODE-BREAK.

IV. WITH A STRONG CURRENT, EXCITATION OCCURS AT KATHODE-MAKE, AT KATHODE-BREAK, AT ANODE-MAKE, AND AT ANODE-BREAK.

The effects of Polar Excitation observed in the animal nerve are thus the same as those observed in the conducting tissue of the plant. In fact it was the observation of reactions of Types III and IV in the plant that led to the discovery of similar reactions in the animal nerve. These characteristic effects, identical in the animal and in the plant throughout a wide range of stimulation, must be regarded as conclusive proof that the nature of the process of conduction of impulse is essentially similar in the two cases.

SUMMARY

It had been ascertained that in the irritable tissue of various sensitive plants excitation is initiated as in animal nerve at kathode-make of a feeble current and is conducted to a distance. With a moderate current, excitation had been found to occur at kathode-make and at anode-break. With still stronger current two other types of reaction III and IV had been observed in the plant.

Parallel experiments were carried out with the sciatic nerves of a toad by the Bipolar Method. Special care was taken to raise the nerves into such position as to remove the possibility of production of secondary poles by contact with neighbouring muscles.

With the same nerve and with identical connections, all the Types I, II, III, and IV were observed, as with the conducting tissue of the plant, on appropriately increasing the intensity of the current.

These results conclusively prove that the transmission of impulse in plants is an excitatory process of the same nature as that of the nervous impulse in animals.

XIV.—ARREST OF CONDUCTION OF IMPULSE BY PHYSIOLOGICAL BLOCK

BY

BINOY K. PALIT, B.Sc.

THE observation of the discriminative polar excitation in plants and of its conduction to a distance affords conclusive proof of the excitatory character of the impulse. Further independent crucial evidence is afforded by the arrest of its transmission by various physiological blocks which are effective in arresting the nervous impulse in the animal. It is obvious that if the transmission were due to a hydro-mechanical disturbance or to the transpiration-current, no arrest by physiological block would be possible.

The nervous impulse in the animal is arrested by the interposition of physiological blocks produced by excessive cold, by local action of poison, and by electrotonus. At the kind suggestion of Sir J. C. Bose I not only repeated the experiments carried out by him but also several additional ones, in order to find whether or not these physiological blocks were effective in arresting the impulse in plants.

THE METHOD OF BLOCK

A strip of cloth 10 mm. in breadth was wrapped round the petiole of Mimosa, midway between the outermost stimulated point and the pulvinus. Two rings of plasticine prevented the possibility of the leakage of the ice-cold water or of the poisonous solution used to produce the block beyond the circumscribed area. The strip of cloth was first moistened with a 2 per cent. solution of glycerine in order to keep the cloth in a moist condition. The normal rate of conduction was first observed, and stimulation by induction shock was applied at the farthest end of the petiole at S. The various blocks were then interposed at B, and the experiment repeated by application of similar testing stimulus. The arrest of the impulse is indicated by the abolition of responsive fall of the leaf on stimulation (fig. 58).

If on the transfer of stimulation to the left of the block at S′ impulse is transmitted, the fact would demonstrate

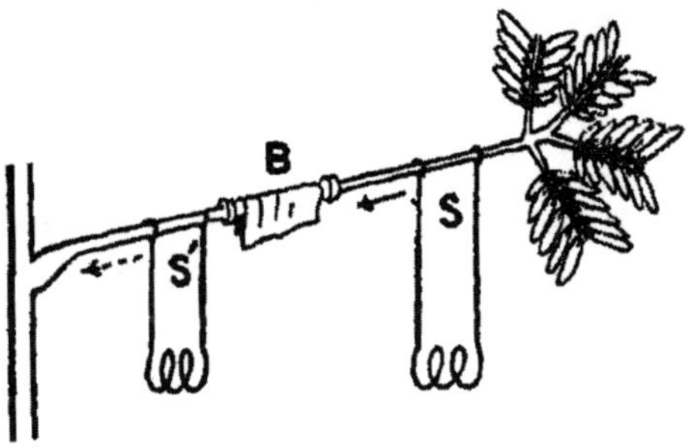

FIG. 58. Diagram of experimental method for determining the effect of physiological block in arresting conduction of impulse in petiole of *Mimosa pudica*.

Stimulation of petiole at S by induction shock; B, block in the path of conduction; S′, stimulation applied to the left of the blocked area.

that the previous arrest of the impulse was due to the physiological block which had been induced locally.

BLOCK OF CONDUCTION BY LOCAL COOLING

The animal nerve, when excessively cooled, loses its power of conduction. A parallel effect is observed in the petiole of Mimosa.

Experiment 77.—The record (fig. 59) gives data for the determination of the velocity of normal conduction. Stimulation was effected at a distance of 20 mm. from the pulvinus. The time-interval between stimulation and response was 1·1 seconds. After making allowance for the latent period of the pulvinus, the average value of which is 0·1 second, the velocity was 20 mm. per second.

Ice-cold water was then applied at *b*, and the subsequent record obtained showed that there was a complete block of

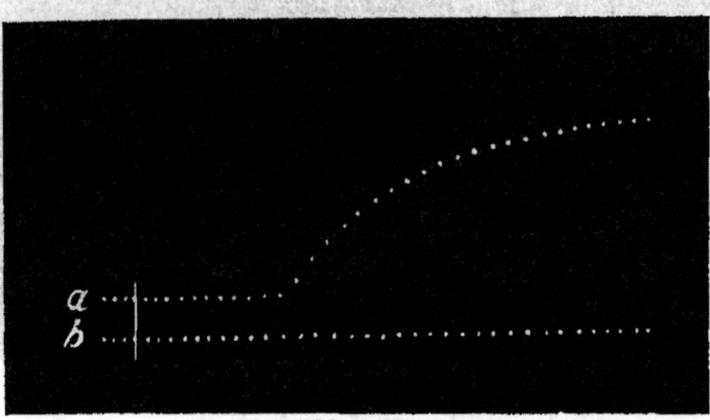

FIG. 59. Ice-cold water acting as a block to the conduction of impulse.

a. Record of normal conduction of impulse.
b. Arrest of conduction by excessive cold (Mimosa).

conduction. This paralysis induced by cold was found to persist for a considerable length of time.

Block of Conduction by Local Action of Poison

Experiment 78.—The application of a poisonous solution on a narrow zone of the petiole of Mimosa was found to induce depression of the power of conduction, which culminated in permanent abolition. KCN solution abolished the conducting power in the course of about 10 minutes.

In the record 1 of fig. 60, the period of transmission of impulse through a length of 26 mm. of the petiole (after allowance for the latent period of the pulvinus) is seen to be 2·0 seconds. The velocity is therefore 13 mm. per second. Application of KCN solution produced a permanent block, as seen in record 2. The leaf was then directly stimulated; it responded after a latent period of 0·1 second, as seen in record 3, showing that the poisonous solution had not affected the moto-excitability of the pulvinus.

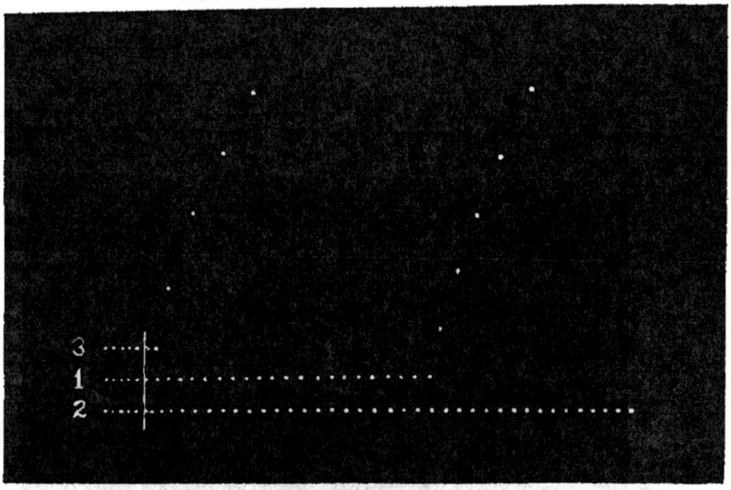

FIG. 60. Abolition of conduction by poisonous solution of KCN.
Record 1. Velocity of normal conduction.
Record 2. Permanent abolition of conduction.
Record 3. Response of pulvinus to direct stimulation (Mimosa)

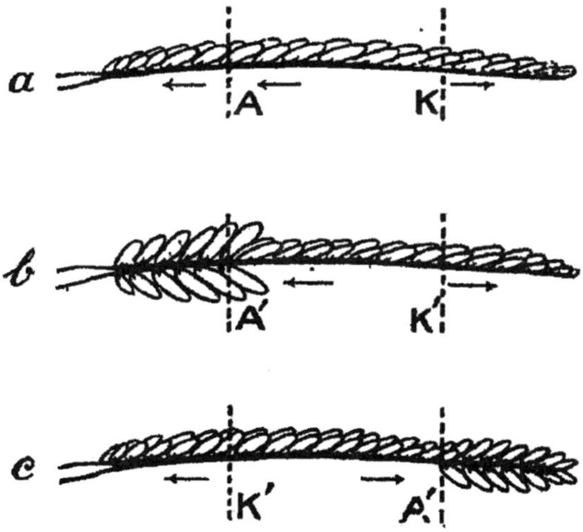

FIG. 61. Block of conduction by strong anode.

In *a* impulse is transmitted across feeble anode; in *b* it is arrested by the stronger anode A′; in *c* the current was reversed and impulse is arrested at the strong anode A′ to the right.

ELECTROTONIC BLOCK

A very characteristic effect in the animal nerve is the arrest of the impulse by the depressing action of a moderately strong anode. A remarkably similar arrest of conduction of impulse occurs in the conducting tissue of the plant as the effect of a strong anodic block.

Experiment 79.—This arrest is demonstrated in a striking manner in the following experiments with a sub-petiole of *Mimosa pudica*. Two electrical connections were made with the sub-petiole at K and A. An E.M.F. of 4 volts was at first employed for excitation. The make of kathode initiated excitation at K (there being no excitation on anode-make at A). The impulse initiated at K travelled inwards and outwards and caused successive closure of all the leaflets. It is to be noted that a feeble anode did not cause arrest of the impulse (fig. 61, *a*).

The E.M.F. was then increased to 8 volts, and the excitatory impulse initiated at K′ travelled, as before, in both directions. But when the centripetal impulse reached the stronger anode A′, it could not traverse the block, so that the impulse was arrested (fig. 61, *b*). Reversal of current caused arrest of the impulse at the strong anode A′ to the right (fig. 61, *c*).

Experiment 80. *Anodic block in sub-petiole of Tree Mimosa.*—A modification was made of the method of stimulation. A tetanising induction shock of intensity of 0·75 unit was applied at S. The excitatory impulse was transmitted inwards till it reached the secondary pulvinus of the sub-petiole, all the leaflets undergoing closure in regular sequence. After the recovery of the leaflets, a polarising electric current was applied from a 20-volt battery, A being made the anode. Application of tetanising shock, of the same intensity as before, started an impulse which was transmitted only up to the strong anode A, which effectively blocked the further passage of the impulse (fig. 62).

The interesting characteristic of the electrotonic arrest

of .mpulse is that it persists only during the passage of the blocking current. Thus, by putting the electrotonic current 'on' and 'off,' the impulse could be alternately

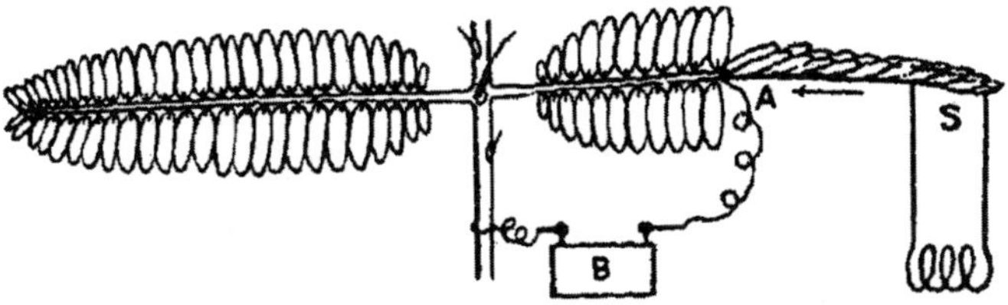

FIG. 62. Impulse generated by tetanising induction shock at S arrested by the strong anode A in sub-petiole of Tree Mimosa. B, battery.

arrested and allowed to proceed without hindrance. The following is an additional experiment on the subject previously described in the 'Nervous Mechanism of Plants.'

EFFECT OF ALTERNATE APPLICATION AND REMOVAL OF ELECTROTONIC BLOCK

Experiment 81.—The experiment was carried out with the petiole of *Mimosa pudica* during spring, when the conducting power was fairly high. A test stimulus of induction shock of 0·3 unit applied at a distance of 30 mm. from the pulvinus was found to be effective. A record was taken of the velocity of transmission by the Resonant Recorder; record 1 (fig. 63) shows that there are 14 spacings between stimulus and response, the time of transmission through 30 mm. being 1·4 seconds. Making allowance of 0·1 second for the latent period of the pulvinus, the velocity was 23·1 mm. per second. An electrotonic block by a current from a 6-volt battery was then produced half-way between the point of stimulation and the pulvinus. During the maintenance of the blocking current, the impulse generated by the testing stimulus was found to be completely arrested, as seen in record 2. After the removal of the blocking current, the record 3 was taken once more under the test stimulus. The block was now found to have been removed with the cessation of the electrotonic current, the velocity of conduction of the impulse being the same as at the

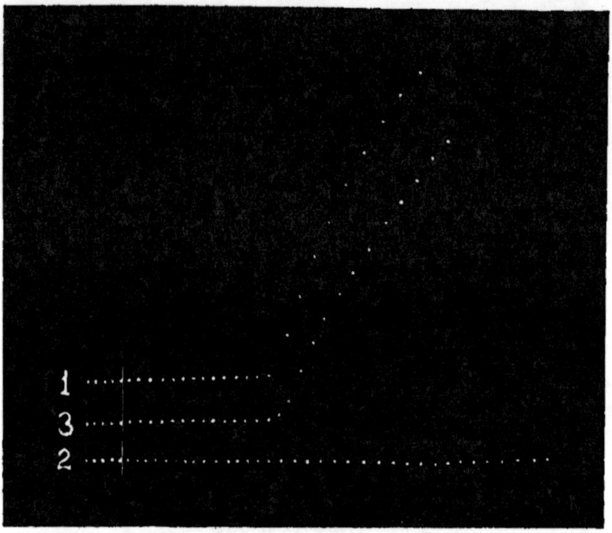

FIG. 63. Arrest of impulse by electrotonic block in petiole of *Mimosa pudica*.

1. Record giving normal velocity of impulse.
2. Arrest of the impulse by the electrotonic block.
3. Restoration of normal velocity of conduction after removal of the block.

Successive dots at intervals of 0·1 second.

beginning. The experiment was repeated with the identical specimen six times in succession with invariable success.

SUMMARY

The various physiological blocks which arrest the transmission of the nervous impulse in the animal, also arrest transmission of impulse in the plant in a similar manner.

Local application of excessive cold paralyses the nerve in the animal as it does the conducting tissue in the plant.

Poisonous solutions cause a permanent abolition of conduction in both.

Transmission of nervous impulse in the animal is arrested by the interposition of an electrotonic block, conduction being restored on the cessation of the electric current. Similarly, transmission of impulse in the plant can be repeatedly arrested or restored by the alternate application and removal of the electrotonic block.

These results afford conclusive proof of the identical nature of the transmission of impulse in the plant and in the animal; in both it is the propagation of protoplasmic excitation.

XV.—THE PARTIAL AND COMPLETE BLOCK OF CONDUCTION OF IMPULSE BY CHLOROFORM

BY

N. N. DAS, M.B., M.SC., AND B. K. DUTT, B.SC.

CRUCIAL evidence in proof of the nervous character of the transmission of impulse in plants has been given in the preceding paper by the action of various physiological blocks, such as those produced by excessive cold, by electrotonus, and by poisonous solutions.

The question has been raised whether an effective block to the passage of the impulse in plants can be produced by a strong narcotic agent. Pfeffer attempted to arrest the impulse in *Mimosa pudica* by chloroform, but failed in his attempt. The conclusion was therefore drawn that the impulse transmitted in a plant under stimulation could not be of an excitatory character as in the animal. It will be presently explained that this inference is erroneous on account of certain defects in Pfeffer's method of experimentation.

BLOCK OF CONDUCTION IN ANIMAL NERVE

The effect of chloroform in producing a block in animal nerve is well known. It is, however, necessary to have definite knowledge of the characteristic effects of chloroform on nerve in inducing partial or complete block, as determined by the strength of the dose and the duration of application. After securing this information, investigation was carried out to find whether, under strictly parallel conditions, partial or complete block was produced in the conducting tissue of the plant.

The method of experiment on animal nerve is as follows: A Frog nerve-and-muscle preparation was made in the usual manner. The nerve was placed on a piece of paraffined cork and the middle portion of it was so bent as to rest in a small conical cup, which, when necessary, could be filled with a very dilute or moderately strong solution of chloroform (fig. 64). The cup was afterwards closed by a plug of cotton-wool.

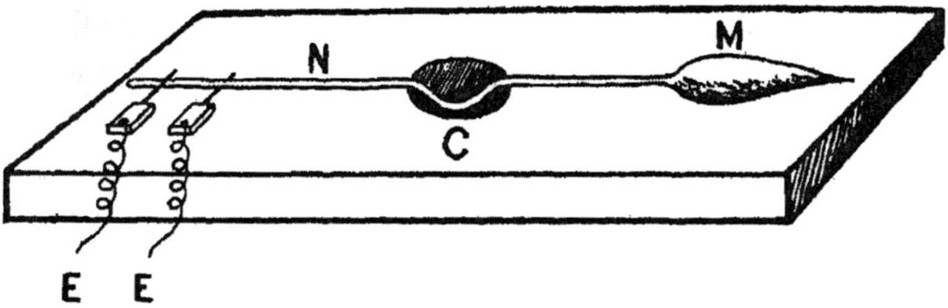

FIG. 64. Nerve-and-muscle preparation, N, M, of Frog for studying the effect of Chloroform on conduction of impulse.

A portion of nerve is placed in conical cup C containing dilute Chloroform; E E, electrodes for stimulation.

Stimulation was effected by a single-break induction shock, the intensity of which could be gradually increased by the approach of the secondary coil towards the primary; the primary coil was actuated by a current of constant intensity. The intervening distance between the secondary and the primary could be varied from 40 cm. to 1 cm. The minimal stimulus which, by its transmitted effect, produced contraction of muscle, was first determined. A few drops of chloroform were then introduced into the cup, in which lay a short length of the nerve. The formerly effective stimulation was now found to be ineffective in the transmission of impulse, proving the depression of conductivity by the narcotic. The block, when partial, could still be forced by increasing the intensity of stimulation by bringing the secondary coil nearer the primary. But when conductivity was abolished, then even the strongest shock failed to induce transmission of impulse through the narcotised area.

Experiment 82. *Effect of very dilute solution of chloroform.*—The minimally effective stimulation was in this case produced when the secondary coil was at a distance of 32 cm. from the primary. Application of dilute chloroform caused a depression of conductivity, shown by partial block of conduction, this stimulation being now found to be ineffective. Conduction of excitation only occurred when the coil was brought nearer—namely, to 30 cm. The trace of chloroform was next removed by washing the nerve, and after 5 minutes the original conductivity was restored, stimulation being once more effective when the secondary was at 32 cm., as at the beginning.

The results are summarised as follows:

Condition of nerve	Effective stimulation at distance between coils
Normal	32 cm.
After dilute chloroform . .	30 ,,
Recovery after removal of narcotic	32 ,,

These results prove:

(1) That dilute chloroform induces a partial block of conductivity.

(2) That the block is relative, and that the partial block can be forced by stronger stimulation.

(3) That there is a restoration of normal conductivity after removal of the trace of chloroform from the nerve.

Experiment 83. *Effect of duration of application of the narcotic.*—A slightly stronger solution of chloroform was now applied to a fresh nerve, and a depression of conductivity was observed under the continued action of the narcotic. The normal transmission occurred when the secondary coil was at a distance of 33 cm. After 1 minute's application of the narcotic, the depression of the conducting power was such that the induced block could only be forced when the secondary was brought to 30 cm.

A greater depression was induced after the application of chloroform for 3 minutes, and the block could now be forced only by bringing the secondary coil to 20 cm. After 5 minutes the depression was still greater, and transmission was only effected under the stronger stimulation produced when the secondary coil was moved to 15 cm.

The following is a summary of results :

Condition of nerve	Effective transmission at distance between coils
Normal	33 cm.
After 1 minute's application of chloroform . . .	30 ,,
,, 3 minutes' application of chloroform . . .	20 ,,
,, 5 minutes' application of chloroform . . .	15 ,,

The effectiveness of the induced block thus depends on the duration of the application of the narcotic.

Experiment 84. *Effect of strong dose of chloroform.*—The normal transmission of impulse occurred when the secondary was at 36 cm. Application of strong solution of chloroform for 1 minute induced a block which could only be forced by bringing the secondary coil to 32 cm.

Condition of nerve	Effective transmission at distance between coils
Normal	36 cm.
After 1 minute's application of narcotic . . .	32 ,,
,, 3 minutes' application of narcotic . . .	19 ,,
,, 5 minutes' application of narcotic . . .	5 ,,
,, 7 minutes' application of narcotic . . .	Abolition of conduction

After 3 minutes the block was not forced until the secondary had been brought to 19 cm. After 5 minutes the coil had to be brought still nearer—namely, to a distance of 5 cm.—for transmission of impulse. After 7 minutes the power of conduction was completely abolished, for the block could not be forced even under the strongest shock.

The results are summarised at foot of previous page.

THE EFFECT OF CHLOROFORM ON CONDUCTION OF IMPULSE IN PLANTS

The manner in which chloroform produces a partial or complete block of conduction in the animal nerve being now clearly understood, parallel experiments were carried out with the conducting tissue of the plant.

Pfeffer, as previously stated, failed to arrest the impulse in Mimosa by the action of chloroform, the failure being due to his employment of exceptionally strong stimulation. The complications and errors which arise from the employment of very intense stimulation have been pointed out by Sir J. C. Bose in pp. 10–12 of his 'Nervous Mechanism of Plants' (1926), from which the following is a quotation:

'It is doubtful whether the conducting tissue in the interior (of the plant) can be effectively narcotised by the external application of an anaesthetic. It is conceivable that after a very long application, a small quantity of the narcotic might, by absorption, get across to the internal conducting tissue; but narcotisation in such circumstances could only be partial, and this partial block might fail to arrest the impulse due to intense stimulation. . . . The most fruitful source of error in the investigation of the conduction of excitation is the employment of drastic modes of stimulation. Though the normal transmission, under the action of moderate stimulus, is confined to the well-defined conducting tissue (contained in the vascular bundle), the excitation becomes diffused under intense stimulation. The same thing also occurs in the nervous conduction in the animal, where under abnormally powerful stimulus the excitation becomes widely diffused. The necessity for discarding crude and drastic methods of stimulation in researches on conduction will now have become obvious. The object of inquiry is not to find out that a violent disturbance becomes diffused, but to determine the nature of the transmission of impulse under normal modes of stimulation. By employing stimulus of graded intensity, it should be easy to determine the character of the impulse generated by observing the effects of various physiological blocks in modifying the power of conduction.'

Bearing in mind the importance of the employment of moderate intensity of stimulation, which has to be maintained constant in successive experiments, two different means of minimally effective quantitative stimulation were employed:

(1) By tetanising induction shock.
(2) By polar action of a constant current.

Block of Conduction in Mimosa

The method of block by interposition of chemical substances has been described in a previous paper. The experiment on the action of chloroform is carried out as follows:

The normal rate of conduction was first observed, the stimulation being due either to tetanising induction shock or to the polar action of a constant current. Having observed the normal rate, the experiment was repeated after application of dilute chloroform on a strip of cloth placed on the plant in the path of conduction (see fig. 66).

Stimulation by Tetanising Induction Shock

The experiment was carried out with a potted *Mimosa pudica*, which was brought into the well-lighted experimental room. When thus transferred from the open air, the excitability of the plant became depressed. The normal excitability was, however, fully restored after a period of 2 hours in the experimental room. The time chosen for the experiment was mid-day, when both light and temperature remained practically uniform. The temperature of the room was 22° C. The specimen was in only moderately vigorous condition.

Before giving detailed data on the action of chloroform, a few words may be said in regard to the effect in relation to the strength of the application. Thus in a particular case the time required for conduction through 25 mm. of the petiole was 2·8 seconds. Application of dilute chloroform caused a great depression of the rate of conduction,

FIG. 65. Block of conduction of impulse in the petiole of *Mimosa pudica* by Chloroform.

Record 1. Time for normal conduction through 30 mm.
Record 2. Block induced by application of chloroform in the path of conduction.
Record 3. Time for conduction through 7·5 mm. of the petiole on the left side of the block.

the time taken for conduction of the impulse being increased from 2·8 seconds to 120 seconds. A stronger dose of chloroform abolished the conducting power. This is well shown in the record given of the next experiment.

Experiment 85.—The velocity of normal conduction in the petiole was obtained from the dotted record given by the Resonant Recorder, the successive dots being at intervals of 0·1 second. Stimulation was first produced at a distance of 30 mm. from the pulvinus. In record 1 there were 31 spacings, each representing 0·1 second (fig. 65). Making allowance for the latent period of the pulvinus, the average value of which is 0·1 second, the true time for transmission through 30 mm. is 3·0 seconds; the velocity of normal conduction is therefore 10 mm. per second. Dilute chloroform was next applied in the path of conduction, and record 2 was taken under similar stimulation as at the beginning; the record shows that the impulse was arrested. In order to prove that this abolition of conduction occurred only in the locally chloroformed area, and that the absence of response was not due in any way to depression of the moto-excitability of the pulvinus, the application of the tetanising induction shock was transferred to the left of the narcotised area, at a distance of 7·5 mm. from the pulvinus. Record 3 shows that the conduction here was normal as exhibited by the fall of the leaf, which occurred after an interval of 0·85 second. The velocity of transmission in the unaffected area was 10 mm. per second, which is the same as that for conduction through the longer path at the beginning. The results are given in the following tabular statement:

(1) Time for normal conduction through 30 mm. . .	3·2 seconds
Latent period of pulvinus .	0·1 second
Velocity of transmission through the long path .	10·0 mm. per second
(2) Conduction became blocked by chloroform.	
(3) Time of conduction through unaffected shorter length (7·5 mm.) . . .	0·8 second
Velocity of conduction . .	10·0 mm. per second

It is thus found:

(*a*) That application of chloroform arrests the impulse.
(*b*) That the abolition of conduction is localised in the anaesthetised region of the petiole.
(*c*) That the conduction is normal in the rest of the petiole not treated with chloroform.

Stimulation by Polar Action of Electric Current

The Bipolar Method of experiment has already been described (p. 113). Suitable electrical connections were made at R with the petiole of the right leaf, and at L with that of the left leaf. Stimulation of R on the kathode-make gives rise to an excitatory impulse which, reaching the pulvinus, causes the fall of the right leaf. Stimulation is then transferred to L by reversing the current, when the left leaf exhibits a responsive fall. The left leaf is in

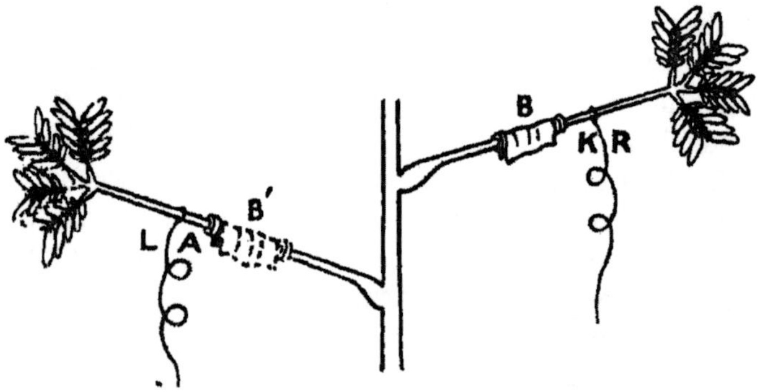

FIG. 66. Effect of Chloroform in blocking conduction in the petiole of Mimosa under polar excitation with constant current.
Impulse initiated on kathode-make at R or at L. Chloroform is locally applied at B on the right side, or on both sides B and B′. A, anode; K, kathode.

some cases employed as a control. A short length of the petiole B on the right, as already explained, is subjected to the action of chloroform. In other cases, as in Experiment 87, chloroform is also applied at B′ on the left petiole (fig. 66).

Experiment 86.—An E.M.F. of 3 volts was found effective to initiate the excitatory impulse at kathode-make. When R was made kathode the impulse was conducted to the pulvinus after an interval of 4 seconds, the distance of transmission being 30 mm.; the velocity of conduction was therefore 7·7 mm. per second. On reversal of the current, excitation was effected on kathode-make at L, the left leaf undergoing a fall after an interval of 3·9 seconds.

On application of dilute chloroform for 10 minutes at B on the right petiole, the impulse initiated at R failed to be conducted across the anaesthetised area, shown by the right leaf not exhibiting any responsive fall. That the normal conducting power had remained unaffected during the period of the experiment was proved by transferring the stimulation to L, effected by reversing the direction of the current. The conducting power of the petiole, used as a control, was unchanged, the fall of the left leaf occurring after an interval which was practically the same as at the beginning of the series of observation. The following is a tabular statement of results :

Kathodic Stimulation of the Right Leaf at R

Time of transmission through 30 mm. 4 seconds
Velocity of conduction . . . 7·7 mm. per second

Conduction arrested after the application of chloroform for 10 minutes.

Control Experiment : Kathodic Stimulation of Left Leaf at L

Time of transmission through 30 mm. 3·9 seconds
Velocity of conduction . . . 8 mm. per second

The velocity in the left petiole not treated with chloroform remained unchanged.

The next experiment was carried out at Hazaribagh in Behar; the leaves were very large, the petiole being more than 60 mm. in length. The irritability of the plant, as previously stated, was still considerable, even at a temperature as low as 15° C.

Experiment 87.—Effective stimulation was induced by an E.M.F. of 4·5 volts. When R was made the kathode, the excitatory impulse travelled through a distance of 60 mm. and caused the fall of the leaf after 10 seconds, the velocity of transmission being about 6 mm. per second. The conducting power, it should be remembered, is low in the winter months. When the stimulation was transferred to L, the fall of the left leaf in response to impulse conducted through 60 mm. occurred after 12 seconds, the velocity being 5 mm. per second. Chloroform was applied on a 10-mm. length of the right petiole at B. The impulse in the right petiole was shown to be arrested, as the right leaf did not show any responsive fall. But transfer of stimulation to L caused the fall of the left leaf as usual.

Chloroform was then applied at B′ on the left petiole. The excitatory impulse initiated at L was now found to be arrested by the block produced by chloroform at B′, just as it had been by the block at B on the right petiole.

In order to prove that the moto-excitability of the pulvinus had not undergone any change during the period of the experiment, an electric induction shock was passed from the point R on the right to L on the left petiole. The unchanged moto-excitability of the two pulvini was exhibited by the fall of both the leaves. The results are summarised in the following tabular statement:

Stimulation of the Right Leaf at R

Time of conduction through 60 mm.	10 seconds
Velocity of conduction . . .	6 mm. per second

Stimulation of the Left Leaf at L

Time of conduction through 60 mm.	12 seconds
Velocity of conduction . . .	5 mm. per second

Application of Chloroform on the Right Petiole

Block of conduction in the right and not in the left petiole.

Application of Chloroform also on the Left Petiole

Block of conduction in both petioles.

The moto-excitability of the two pulvini remained unchanged.

SUMMARY

A crucial proof of the excitatory character of an impulse is that it can be blocked by application of poison in the path of conduction. The permanent abolition of conduction of the impulse in plants through an area poisoned by the action of potassium cyanide has already been demonstrated.[1]

The effect of chloroform in arresting the impulse in animal and in plant has been investigated under parallel conditions.

It has been shown (1) that in the animal nerve dilute chloroform induces a partial block of conductivity; (2) that the partial block can be forced by stronger stimulation; (3) that prolonged application of strong chloroform induces a permanent abolition of conductivity.

The foregoing experiments demonstrate that the conduction of impulse in the petiole of *Mimosa pudica* is affected by the action of chloroform in exactly the same way as is the conduction of nervous impulse in the animal.

Pfeffer's failure to produce a block of conduction in Mimosa is to be attributed to (1) imperfect anaesthetisation by chloroform; (2) the employment of too intense stimulation.

[1] Bose, *Irritability of Plants* (1913), p. 173.

XVI.—EFFECT OF CHANGE OF ENVIRONMENT IN MODIFYING THE INTERNAL PHYSIOLOGICAL ACTIVITY OF THE PLANT

BY

SURENDRA CHANDRA DAS, M.A.

THERE are two factors which determine the nature of the characteristic life-reactions of a plant; one of these is its inherent properties, and the other is the changes induced by variation in the environment. In what way would it be possible to estimate the internal activities of the plant? Certain aspects of irritability have been observed and measured, so that a sort of standard value has here been reached. By observing how these functions are affected by change in environment, an indication is obtained of the effect of change on the general physiological activity of the plant.

The difficulties which surround the investigation of the problem have been successfully removed by various devices perfected in the Institute, which enable the plant to record automatically its response to a testing shock and thus reveal its internal condition.

There are four aspects of irritability of the plant which are of special importance. These are: (1) the *Moto-excitability*, by which a contractile movement of the organ is effected on stimulation; (2) the *Latent Period*, or relative quickness of the organ to respond to stimulation; (3) the *Apex-Time*, or the time required for the organ to attain its maximum contraction after the incidence of stimulus; and (4) the *Conductivity*, or the power of conducting an excitatory impulse to a distance.

Sir J. C. Bose's kind suggestions have greatly encouraged me to study the effect of changes of environment in modifying the internal activities of plants. I am much indebted to him for valuable advice received from time to time.

METHOD OF EXPERIMENTATION

In order to determine the internal variation induced by the environment, two batches of *Mimosa pudica* plants were grown under different conditions—namely, under the open sky and within a semi-shaded greenhouse. When Mimosa is grown in the open, the upper side of the petiole becomes reddish-brown in contrast to the opposite or under side, which is greenish in colour. There is, as it were, a sharp line of demarcation separating the upper and under halves of the petiole and pulvinus. The leaflets in these specimens are dark green on the upper and light green on the under surface.

When a batch of plants was transferred to the semi-shaded greenhouse, it was observed that the old leaves began to wither and fall off in the course of three or four days. New leaves, however, commenced to sprout out which were light green; the petioles of these plants were thinner and longer than those of plants growing in the open. These leaves belonged, as it were, to a new generation tuned to a different environment.

So far in regard to visible external change. What of the invisible internal variation? Could it be detected by the automatic record given by the plant to a testing stimulus?

The various physiological activities which might be expected to undergo a marked variation are:

(1) Moto-excitability.
(2) Latent Period.
(3) Apex-Time.
(4) Conductivity.

The Resonant Recorder

The automatic records of the plant were taken by means of the Resonant Recorder which has been fully described elsewhere.[1] The writer is a fine steel wire which has to be maintained in a state of resonant vibration, being exactly

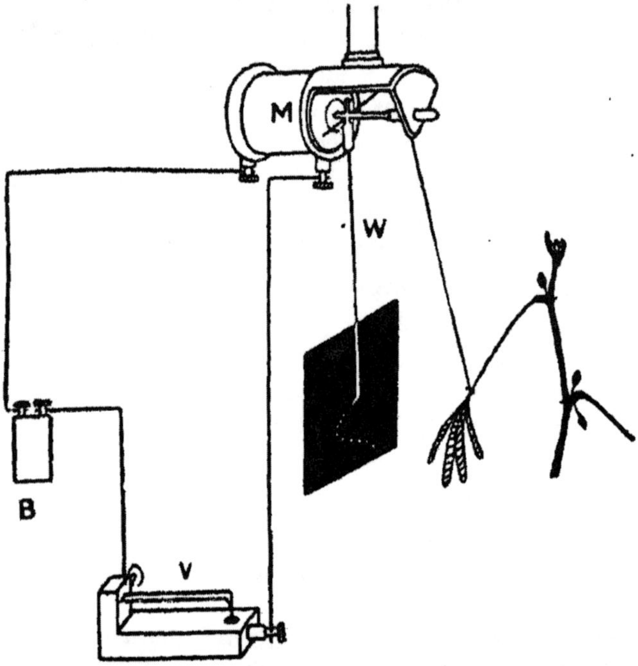

Fig. 67. Diagrammatic representation of Resonant Recorder. M, electro-magnet; W, resonant writer; V, vibrating reed; B, battery.

tuned to vibrate, say, ten times in a second; it is supported on jewel-bearings at the centre of one pole of a circular electro-magnet. The magnetising coil is in circuit with a storage cell B and a vibrating reed V, which periodically comp etes the electric circuit (fig. 67). When the reed is exactly tuned to vibrate ten times in a second, the recording writer is thrown

[1] Bose, *Irritability of Plants* (1913).

into sympathetic vibration and strikes the smoked-glass plate once every tenth of a second. With finer recorders it is possible to measure time-intervals as short as a hundredth or a thousandth of a second. With a writer vibrating ten times in a second, the successive dots in the record measure time-intervals of 0·1 second. The advantage of a dotted record is (1) that the error arising from friction is removed, and (2) that the phytogram is its own chronogram.

The experiments were carried out between the hours of 11 A.M. and 3 P.M. in a room near a window admitting northern light of the sky. The temperature and light were practically uniform during those hours.

MOTO-EXCITABILITY

The moto-excitability of the pulvinus of Mimosa is indicated by the amplitude of response of the leaf to a definite stimulus. Increased excitability is exhibited by a greater amplitude of the response.

Experiment 88. *Moto-excitability of the pulvinus of plant growing in the open.*—A specimen of *Mimosa pudica* grown in the open was taken. Connections from the secondary coil of an induction apparatus were made, one on the petiole and the other on the stem a little below the pulvinus which could thus be directly stimulated. In this and in the following experiment the testing stimulus was an induction shock kept constant at 0·75. The amplitude of contractile response shown in fig. 68, *a*, is comparatively large, being 69 mm. It is to be noted that the magnification produced by the recording lever was the same throughout the following series of experiments.

Experiment 89. *Moto-excitability of the pulvinus of plant grown in greenhouse.*—The response to the testing stimulus was recorded under parallel conditions (fig. 68, *b*). Comparison with fig. 68, *a*, reveals a marked difference in excitability of the plants grown in the open and those grown in a semi-shaded greenhouse.

The typical responses of the plants grown in the open, *a*, and of those grown in the greenhouse, *b*, though taken separately, are, for facility of comparison, printed one above the other in most of the records.

Referring to fig. 68, *a*, the vigorous condition of the plant grown in the open is demonstrated not only by the very large amplitude of response, but also by the relatively greater distance between the successive vertical dots. The

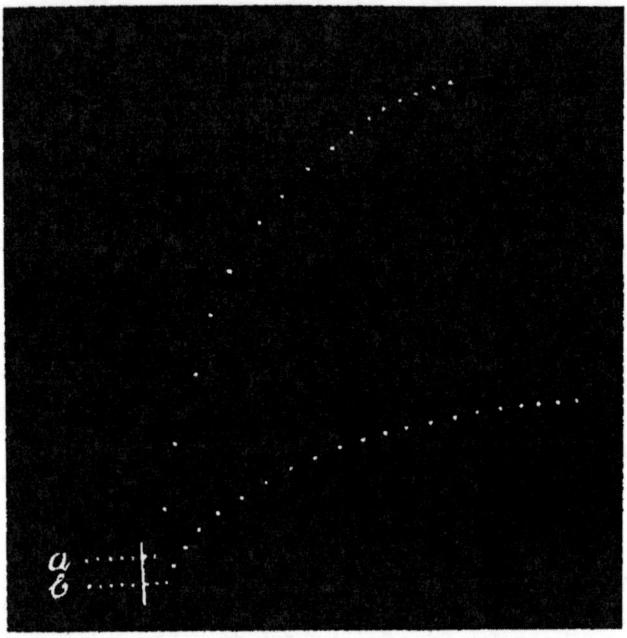

FIG. 68. Records of contractile response (*a*) of Mimosa grown in the open, and (*b*) of Mimosa grown in the greenhouse.

Note short latent period and larger amplitude of response in *a*, and longer latent period and feebler amplitude of response in *b*. Successive dots at intervals of 0·1 second.

curve of record of the greenhouse plant, *b*, is comparatively flat and the amplitude of response in this case is only 23 mm., *i.e.* one-third of that given by the plant grown in the open.

Tables XIV and XV give the data for determining the moto-excitability in four typical plants of each of the two batches, grown in the open and in the semi-shaded greenhouse respectively.

TABLE XIV.—MOTO-EXCITABILITY OF FOUR TYPICAL PLANTS GROWN IN THE OPEN.

Number of specimen	Amplitude of response of leaf in mm.
	66
	69
	72
	74
The average amplitude .	70·25

The average amplitude of the four plants grown in the open is therefore 70·25 mm.

TABLE XV.—MOTO-EXCITABILITY OF FOUR TYPICAL PLANTS GROWN IN THE GREENHOUSE.

Number of specimen	Amplitude of response of leaf in mm.
1	10
2	10·5
3	23
4	24
The average amplitude . . .	16·9

The average amplitude for the plants grown in the greenhouse is 16·9 mm. The moto-excitability of Mimosa grown in the greenhouse was therefore depressed to about one-fourth as compared with the normal excitability of the plant grown in the open.

THE LATENT PERIOD

When the pulvinus of Mimosa is directly stimulated by an electric shock from an induction coil, a brief period elapses between the incidence of stimulus and the initiation of the responsive contractile movement. This 'lost time' is termed the Latent Period.

Experiment 90. *Determination of Latent Period in plants grown in the open and in the greenhouse.*—In order to determine the latent period, the pulvinus was directly stimulated by the passage of an induction shock through it, one electrical connection being made with the petiole and the other with the stem, the pulvinus being interposed between the two. The steel writer employed for the

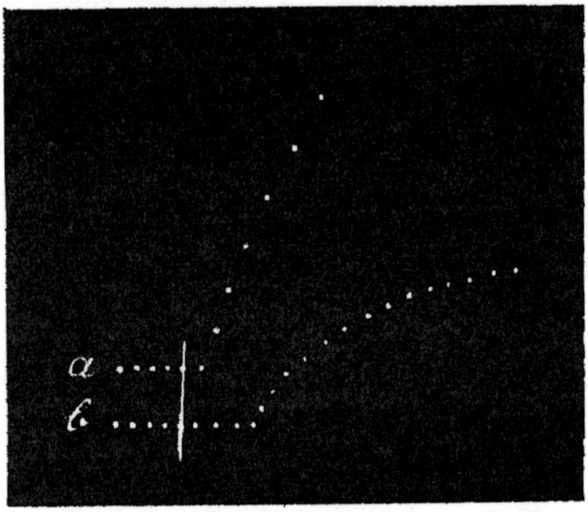

FIG. 69. Records giving the latent period (*a*) of Mimosa grown in the open, and (*b*) of Mimosa grown in the greenhouse.

Note that in *b* the latent period is four times longer than that in *a*. Successive dots at intervals of 0·05 second.

following experiments was tuned to vibrate twenty times per second, so that the interval between successive dots was 0·05 second.

Figure 69, *a* (upper curve), is the typical record of the latent period a plant grown in the open, which in this particular case was 0·05 second; the vigorous contraction of the pulvinus is shown by the large vertical distance between the successive dots.

In striking contrast with the short latent period of Mimosa grown in the open, is the comparatively long latent period of the subtonic plant grown in the greenhouse. The latent period (fig. 69, *b*) is here as long as 0·2 second, that is to say, it is four times as long as that of the highly active specimens grown in the open. The nearly horizontal slope of the curve seen in the lower record also demonstrates the depressed contractility of the greenhouse plant.

The two following tables give the latent period in four typical cases of two batches of Mimosa grown in the open and in the greenhouse.

TABLE XVI.—THE LATENT PERIOD IN PLANTS GROWN IN THE OPEN.

Number of specimen	Latent period in se
1	
2	
3	
4	0·10
Mean value	0·07

TABLE XVII.—THE LATENT PERIOD IN PLANTS GROWN IN THE GREENHOUSE.

Number of specimen	Latent period in se
1	0·15
2	0·20
3	0·20
4	0·20
Mean value .	

The average value of the latent period in plants grown in the greenhouse is therefore 2·7 times as long as that of plants grown in the open.

THE APEX-TIME

This convenient term will be used to indicate the time required for the attainment of maximum contractile fall of the leaf after the incidence of the stimulus. The records given in fig. 70 clearly show the characteristic difference between the apex-time of a plant grown in the open, *a*, and that of a plant grown in the greenhouse, *b*. The

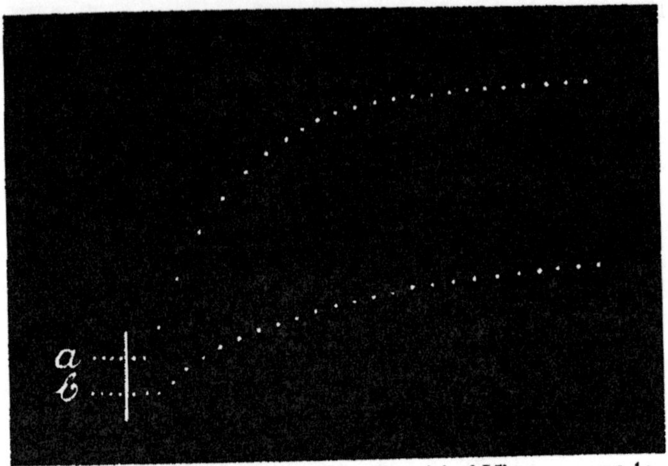

FIG. 70. Records giving the apex-time (*a*) of Mimosa grown in the open, and (*b*) of Mimosa grown in the greenhouse.

Note that while the apex-time in *a* is 1·3 seconds, in *b* it is 2·5 seconds. Successive dots in this and in the following records are at intervals of 0·1 second.

successive dots in the record indicate time-intervals of 0·1 second.

In *a* the latent period is 0·1 second; the contractile reaction then increases rapidly, the maximum rate being attained at the fourth spacing, *i.e.* 0·4 second, after the application of stimulus. The rate then undergoes a diminution, the maximum contraction being attained at the thirteenth dot after initiation of response; the apex-time is therefore 1·3 seconds.

In the case of the plant grown in the greenhouse the characteristics of the response are very different. The latent period is prolonged to 0·2 second; the contractile reaction increases at a very slow rate as indicated by the flat curve, the maximum contraction not having been attained until after the twenty-fifth dot. The apex-time of the greenhouse plant is therefore 2·5 seconds, or nearly twice as long as that of the more vigorous plant grown in the open.

Tables XVIII and XIX give the values of the period of maximum contraction in four typical cases of each batch of Mimosa plants.

TABLE XVIII.—THE APEX-TIME OR TIME OF MAXIMUM CONTRACTION IN PLANTS GROWN IN THE OPEN.

Number of specimen	Period of maximum contraction
1	1·1 seconds
2	
3	
4	1·3
The average value	1·26 seconds

TABLE XIX.—THE APEX-TIME OR TIME OF MAXIMUM CONTRACTION IN PLANTS GROWN IN THE GREENHOUSE.

Number of specimen	Period of maximum contraction
1	1·8 seconds
2	2·0 ,,
3	2·0 ,,
4	2·5 ,,
The average value . .	2·1 seconds

The average value of the time of maximum contraction of the plant grown in the greenhouse is therefore 1·7 times as long as that of the plant grown in the open.

The Conductivity

The velocity of transmission of impulse was determined by the application of electric stimulation on the petiole at a definite distance, *d*, from the pulvinus, the time-interval T between stimulation and response being indicated by the dotted record, the successive dots in which are at intervals of 0·1 second. In order to determine the true time of transmission *t*, it is necessary to subtract the latent period L from the observed time T. The velocity of transmission is found by dividing the distance *d* by the time *t*.

$$V = \frac{d}{t} = \frac{d}{T - L}.$$

Experiment 91. *Velocity of transmission in the petiole of* Mimosa pudica *grown in the open.*—The intensity of the

FIG. 71. Determination of velocity of transmission of impulse in Mimosa grown in the open.

Lower record T gives the interval between stimulation and response; L is the latent period of the pulvinus.

electric stimulus was 1 unit applied for 0·1 second on the petiole at a distance of 30 mm. from the pulvinus. The contractile response occurred 1 second after the stimulation (lower record, fig. 71), and the leaf recovered after

15 minutes. In order to determine the latent period L of the pulvinus, it was subjected to direct stimulation. The upper curve of fig. 68 shows that the latent period was 0·1 second. The true time of transmission of the excitatory impulse through the intervening distance of 30 mm. from the stimulated point on the petiole to the responding pulvinus is therefore

$$0 \cdot 1 = 0 \cdot 9 \text{ second}$$

$$d = 30 \text{ mm.}$$

$$V = \frac{d}{t} = \frac{30}{0 \cdot 9} \quad \text{mm. per second,}$$

where V is the velocity of transmission of the excitatory impulse in this particular petiole of Mimosa. The temperature of the room during the experiment was 29° C.

Below are given the data for the determination of the velocity of conduction of excitation in four typical plants grown in the open.

TABLE XX.—VELOCITY OF TRANSMISSION OF IMPULSE IN PLANTS GROWN IN THE OPEN.

Number of specimen	Distance in mm.	Total time of transmission T in second	Latent Period L in second	True time t in second	Velocity of transmission in mm. per second
1	30	1·0	0·1	0·9	33·3
2	10	0·5	0·1	0·4	25·0
3	15	0·7	0·1	0·6	25·0
4	15	0·8	0·1	0·7	21·4

Average velocity of transmission . 26·2 mm. per second

Experiment 92. *Determination of velocity of transmission in the petiole of Mimosa grown in the greenhouse.*—The intensity and duration of stimulation were the same as in the last case. The striking feature in the reaction of the specimen grown in the greenhouse is the very slow velocity of its conduction and the prolongation of its latent period. Stimulation was applied at a distance of 10 mm., and the lower record (fig. 72) shows that the time of transmission T

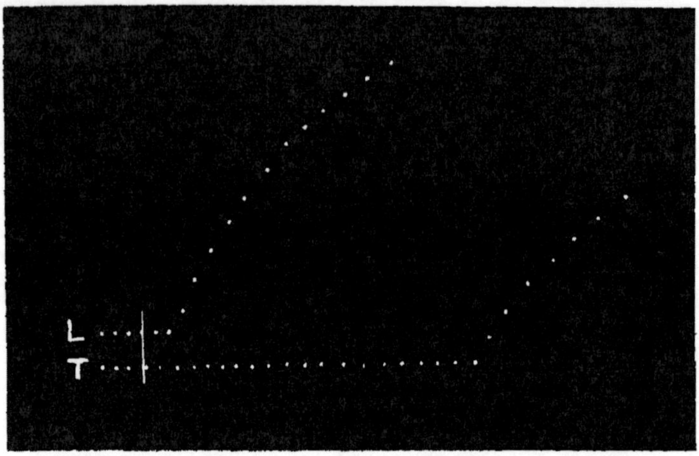

FIG. 72. Determination of velocity of transmission of impulse in Mimosa grown in the greenhouse.
T gives time-interval between stimulation and response; L the latent period of the pulvinus.

was equal to 1·9 seconds. The latent period as determined by the upper record was 0·2 second:

$$t = 1{\cdot}9 - 0{\cdot}2 = 1{\cdot}7 \text{ second}$$

$$V = \frac{d}{t} = \frac{10}{1{\cdot}7} = 5{\cdot}9 \text{ mm. per second}$$

Below are given data for determination of the velocity of transmission of impulse in four typical plants grown in the greenhouse.

The ratio of average velocity of transmission in the two cases is $\frac{26{\cdot}2}{7} = 3{\cdot}75$—that is to say, under less favourable conditions in the greenhouse the velocity of transmission is reduced to nearly one-fourth the normal rate.

TABLE XXI.—VELOCITY OF TRANSMISSION OF IMPULSE IN MIMOSA GROWN IN THE GREENHOUSE.

Number of specimen	Distance in mm.	Total time of transmission T in seconds	Latent Period L in second	True time t in seconds	Velocity of transmission in mm. per second
1	10	2·00	0·20	1·80	5·5
2	10	1·90	0·20		5·9
3	10	1·50	0·20		
4	10	1·25	0·15	1·10	
Average velocity of transmission of impulse				7 mm. per second	

SUMMARY

Variation of the environment induces marked and definite effects in the internal activities of the plant.

Some of these changes can be detected by means of the automatically recorded response given by the plant to a definite testing stimulus.

Changes which are capable of exact measurement are: those in (1) the Moto-excitability; (2) the Latent Period; (3) the Apex-Time; and (4) the Conductivity.

In order to induce the environmental effects, two similar batches of plants of *Mimosa pudica* were grown, the one in the open and the other in a slightly shaded greenhouse. This difference of conditions was found to have induced marked differences in the physiological activities of the two batches.

The moto-excitability of a typical plant grown in the open, as measured by the amplitude of its contractile response, was found to be nearly four times as great as that of a plant grown in the greenhouse.

The plant grown in the open reacted to stimulation about three times more quickly than the greenhouse plant. The average latent period of the former was found to be

0·07 second, whereas it was 0·19 second in the case of the specimen grown in the greenhouse.

The maximum contraction was attained twice as quickly in plants of the batch grown in the open as in those grown in the greenhouse. The average apex-time of a plant grown in the open was 1·25 seconds, while of that grown in the semi-shaded greenhouse it was 2·1 seconds.

The conductivity was also considerably greater in the first than in the second batch of plants. The average velocity of transmission of impulse in plants grown in the open was 26·2 mm. per second, while the average velocity in plants grown in the greenhouse was 7 mm. per second, or about one-fourth the normal rate.

By the methods adopted, it is possible to measure the effect of different environments upon certain functions of the plant, and so to obtain some definite idea of the effect of such differences of environment upon its general physiological activity.

XVII.—EFFECT OF UNILATERAL ACTION OF LIGHT ON OSCILLATING ORGANS

BY

GURU PRASANNA DAS, L.M.S.

THE phototropic effect of light on longitudinally growing organs has been described in 'Growth and Tropic Movements of Plants.' It has been there shown[1] that under the unilateral action of light,

(1) The directly stimulated proximal side undergoes contraction, whereas the opposite distal side undergoes expansion. The conjoint effect of these two factors is to induce a positive curvature of the organ towards the light.

(2) Similar phototropic responses are given by pulvinated and by growing organs.

(3) In growing organs the tropic curvature is brought about by the differential growth of the proximal and the distal sides.

(4) The variation induced under the stimulus of light is modified by the tonic condition of the organ. In a condition of subtonicity, the response of the organ is of opposite sign to the normal—an acceleration of growth instead of the normal retardation.

(5) In certain organs which possess transverse conductivity, conduction of excitation to the distal side causes neutralisation or actual reversal of response from the normal positive curvature towards light to negative curvature away from it.

[1] Bose, *Growth and Tropic Movements of Plants* (1929), p. 101.

In order to determine the specific action of light in modifying growth, it is essential that specimens should be employed which have been found to be extremely sensitive to photic stimulation, so that record can be taken within a short time; otherwise changes in the environment would introduce many complications. It is not, however, easy to secure for the experiments suitable specimens sufficiently sensitive to light; for an organ may prove to be very sensitive to light at certain months of the year, but insensitive at others. The tonic condition on which the character of response depends is also subject to seasonal variation.

The following results relate to the action of diffused light in causing variation in the rate of growth, and of the unilateral action of light in inducing tropic curvature.

The effect of diffused light on growth can be demonstrated by the highly sensitive method of record taken under exact balance. Exposure to light is found to upset the balance, which indicates that the rate of growth is undergoing retardation. This is true of organs which are in a favourable tonic condition.

Positive Phototropic Curvature

Experiment 93. *Unilateral action of light.*—After the demonstration of the fundamental effect of the action of light in inducing retardation of the rate of growth, to which reference has just been made, it is easy to foresee the effect of the unilateral action of light. The directly stimulated proximal side will exhibit retardation, while the indirectly stimulated distal side will show acceleration of growth, the result being concavity of the proximal and convexity of the distal side. The organ will therefore bend towards the light and exhibit a positive phototropic curvature.

The above inference is fully verified by the two uniform positive responses given by the flower-stalk of Allium (fig. 73). The duration of the unilateral application of light was for 2 minutes; the movement towards the light occurred

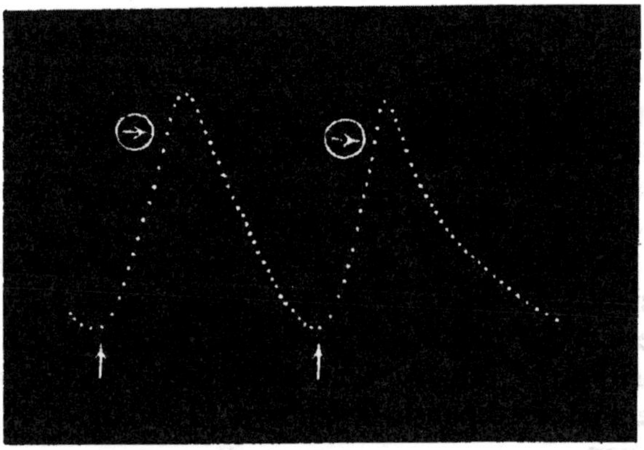

FIG. 73. Record of positive phototropic response of Allium.

Exposure to light began at vertical arrow and ended at arrow within circle. Up-curve represents movement towards light ; down-curve, the recovery.

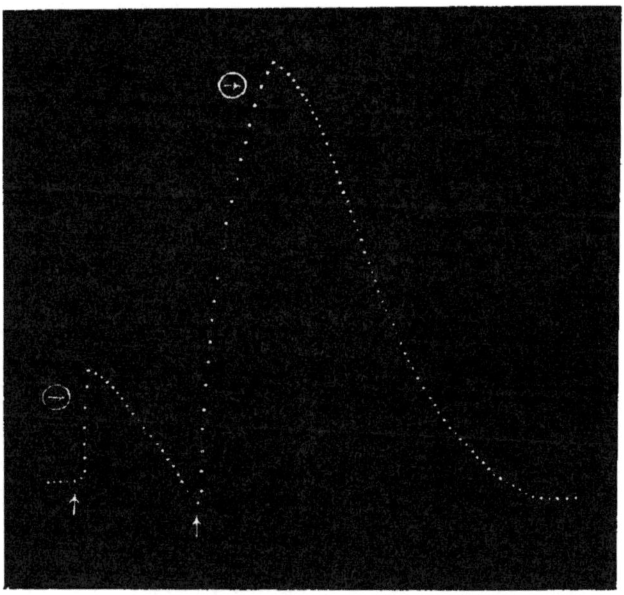

FIG. 74. Effect of duration of exposure in increasing phototropic curvature.

Note increased amplitude of response, and prolongation of period of recovery under longer exposure (Crinum).

not only during the exposure, but persisted for a further period of 40 seconds, the recovery being completed in the course of about 5 minutes.

Experiment 94. *Effect of duration of exposure.*—The tropic curvature is increased under stronger intensity or longer duration of exposure. In fig. 74 are given two records taken with Crinum lily. The source of light was Pointolite of 25 candle-power. The first record is that of an exposure of 5 minutes, and the second that of an exposure of four times as long, *i.e.* 20 minutes. The amplitude of response under prolonged exposure was about four times that under a shorter exposure. The recovery is seen to be protracted after prolonged exposure; the time of recovery from 5 minutes' exposure is 20 minutes, while that after an exposure of 20 minutes is 1 hour.

Modifying Effect of Tonic Condition on Response

Light has been shown to induce retardation of growth in an organ in favourable tonic condition. But in subtonic organs light causes an acceleration of the rate of growth. Stimulation itself is, however, found to raise the tonic level of the organ, so that under continuous stimulation the response is transformed from acceleration to normal retardation.

Experiment 95.—The above conclusion is strikingly demonstrated in an experiment with a subtonic specimen of Wheat. The record was taken by the method of balance, the record being at first horizontal. Diffuse stimulation by light, on exposure at the vertical arrow, caused an upsetting of the balance in an upward direction, indicative of the enhancement of the rate of growth. This acceleration under continued action of light lasted for 6½ minutes, after which it was transformed into normal retardation shown by the downward curve. On the cessation of exposure at the horizontal arrow within the circle, the retardation of growth persisted as the after-effect of light for more than 2 minutes. The rate was then recovered to the normal as represented by the horizontal record of balance (fig. 75). The state of subtonicity is induced by unfavourable conditions of the environment. Thus it was observed that many plants which exhibited the normal retardation of growth under the stimulus of light in spring and summer were brought into a condition of subtonicity

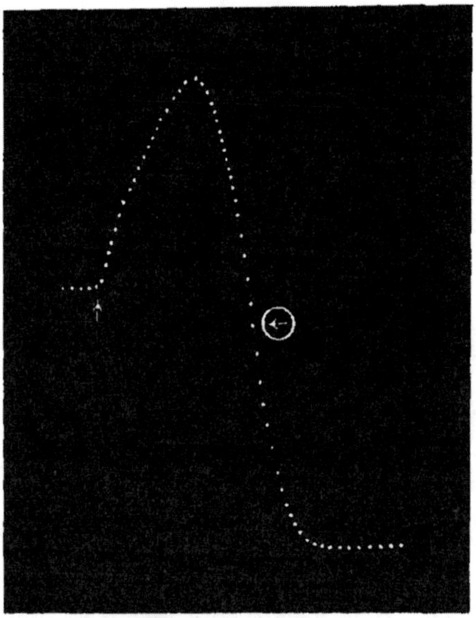

FIG. 75. Effect of light on growth of a subtonic specimen. Record taken by the method of balance. Exposure to diffuse light at vertical arrow enhances rate of growth, shown by up-curve. Continued exposure to light transforms acceleration into retardation, shown by down-curve. On cessation of exposure at horizontal arrow within circle, the induced retardation persisted for 2 minutes followed by restoration of normal rate, shown by horizontal curve (Seedling of Wheat).

by the excessive moisture during the rainy season, so that their response to the stimulus of light was transformed from retardation to acceleration of growth. It is obvious that the phototropic response of subtonic organs would be negative instead of the normal positive.

The experiments on the tropic effect of light which have hitherto been described were made on rigid organs, that is the path of whose growth is a straight line. But it has been discovered that growing organs do not all move vertically upwards, but that some exhibit the most remarkable automatic oscillatory movements, the apex moving alternately to the right and to the left in regular sequence. The object of the following experiments is to ascertain what are the phototropic reactions of such growing organs.

Oscillatory Movements of a Growing Organ

Experiment 96.—Such movements are exhibited, for example, by the internode above the cotyledons in *Vigna Catjang* and in *Vicia Faba*. A record of this automatic

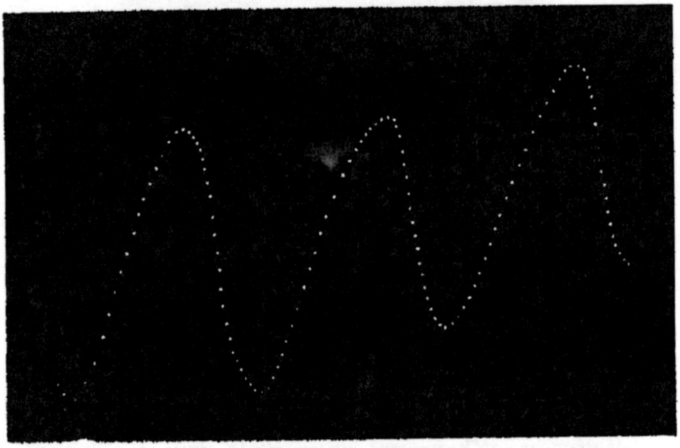

FIG. 76. Record of automatic oscillations of stem of *Vigna Catjang*.
Successive dots at intervals of 1 minute.

oscillatory movement of *V. Catjang* is given in fig. 76, the successive dots representing intervals of 1 minute. The oscillations are seen to be fairly uniform; a movement to the right represented by the up-curve is followed by an opposite movement to the left indicated by the down-curve, the two halves of the oscillation being practically equal. The average period of a single periodic movement is about 41 minutes.

The question naturally arises whether such periodic

movements of growing stems are unique, or whether they are of frequent occurrence. A very similar phenomenon is observed in the torsional movement of twining stems. The autonomous torsion of young circumnutating stems is due to the fact that the rate of growth at any moment is not the same on all the sides; there is a maximum rate which slowly travels round the growing region from segment to segment, so that the apex of the organ describes a circular or elliptical movement round the vertical. The oscillatory movement of a growing stem is only the particular instance in which maximum growth takes place alternately on two opposite sides of the organ. Thus, while one side of the organ is practically in a state of quiescence, the opposite side is in a state of maximum activity; the maximum growth is then transferred to the opposite side, while the previously active side becomes quiescent, and thus oscillatory movements are produced. Had the rate of growth been equal on the two sides, there would have been a more or less uniform longitudinal growth and no oscillatory movement. Other instances are known of oscillatory movement caused by the alternate increase of turgor on the opposite sides of the organ.

Movements of the Lateral Leaflet of Desmodium

A strictly analogous phenomenon to that of the oscillatory movement of the stem is exhibited by the leaflet of Desmodium. The orbit of the movement is often elliptical, due to the wave of variation of turgor that travels round the pulvinule of the leaflet. Under appropriate conditions, however, the movements occur strictly in an up- and down-plane. The upper and the under sides of the pulvinule of the Desmodium leaflet thus correspond to the right and left sides of the oscillating growing stem.

The question next to be determined is whether the tropic response of an oscillating growing or a pulvinated organ to light is essentially similar to that of a rigid, *i.e.* a non-oscillating organ. For the solution of this problem investigations were undertaken on the action of light on both leaflets and stems which were in a state of active movement. I have been encouraged to carry out investigations on the subject at the suggestion of Sir J. C. Bose.

Response of Desmodium Leaflet to Unilateral Action of Light

The chief experimental difficulty lay in the local application of moderately strong light alternately on the upper and the under sides of the very delicate pulvinule. When strong light from an arc lamp was applied, say, from above, it became diffused by reflection from the sides and the floor of the plant-chamber, so that it seemed impossible to subject only one side of the pulvinule to light.

The difficulty was obviated as follows: The record was taken in a room in which there was only feeble diffused light. Arrangements had to be made such that the light incident on either the upper or the under side of the pulvinule should be very much stronger than the feeble light that acted on all sides of the organ. It was also necessary that when strong light was employed, it should act strictly either only on the upper or only on the under side of the thin pulvinule. The most suitable source of light for the present purpose was found in the very small incandescent electric pea-lamp which, with a 4·5-volt battery, gave out very bright and intense light. Though the total candle-power of the pea-lamp was not very great, yet the effective intensity of the incident light could be sufficiently increased by bringing the lamp near to the pulvinule, this being rendered possible by the very small size of the pea-lamp. The light was of moderate intensity when the lamp was at a distance of 10 mm. from the pulvinule; the intensity, however, became greatly increased when the distance was reduced to 5 mm. In regard to the application of light strictly on the upper or on the under side of the organ, the small size of the source of light offered the special advantage that the pulvinule cast a well-defined shadow, on account of which the opposite side was protected from light.

TROPIC EFFECT OF LIGHT ON THE PULVINULE

The method of procedure is to take first the record of normal movement under feeble diffused light. Light from the pea-lamp was then applied either on the upper or on

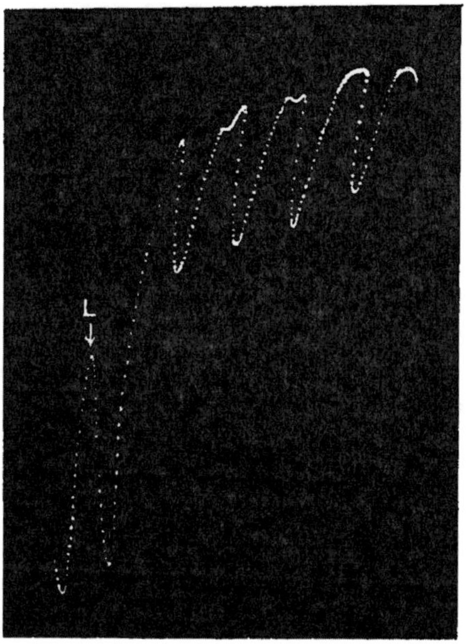

FIG. 77. Phototropic effect of light acting on the upper surface of pulvinule of Desmodium leaflet. Arrow represents beginning of exposure.

Note bending of pulvinule upwards towards light, the upward displacement of the mean line of equilibrium, the inhibition of oscillatory movement downwards away from light, and diminished amplitude of oscillation.

Up-curve represents up-movement, and down-curve downward movement.

the under side of the pulvinule, the change in the record exhibiting the characteristic effect induced by the unilateral action of light.

In the normal record of the oscillation given in fig. 77, the up-curve represents the diastolic up-movement; the down-curve, on the other hand, indicates the systolic down-movement. For facility of description of the effect of light, it would be convenient to use the term *the mean line of equilibrium* for the imaginary line which cuts the oscillation at the middle of the swing, half-way between its two extreme limits.

Experiment 97. *Effect of light acting from above.*—When moderately strong light was applied from above, it caused a displacement of the record in an upward direction (*see* fig. 77). This is due to the bending of the pulvinule upwards by the induced contraction of the directly stimulated upper side. The positive curvature towards the light can be independently verified by direct observation of the attitude of the leaflet. The result of this experiment leads to the conclusion that

The phototropic effect of light on an oscillating organ is essentially the same as that on a rigid growing organ, namely, to induce a positive curvature towards the light.

Careful inspection of the record shows, further, that the extent of oscillation is increased in a direction towards the light, while there is an inhibition of movement in the opposite direction. The result of this is a displacement of the mean line of equilibrium towards the light.

This particular inhibition of the movement away from the light also causes a diminution of the amplitude of oscillation. It may thus happen that under the unilateral action of strong photic stimulation the maximally displaced leaflet towards the light is held there in a fixed position.

Having observed the effect of light acting from above, a further experiment was carried out to determine the effect of light acting from below.

Experiment 98. *Effect of light acting from below.*—This caused a displacement of the record downwards, due to the contraction of the under side of the organ induced by the direct photic stimulation. Here, also, as in the case of light acting from above, a positive curvature is induced towards the light. The mean line of equilibrium is displaced towards the light, this time downwards. The amplitude of the oscillation is also decreased by the inhibition of the natural up-movement away from the light (fig. 78). The results of this and of the previous experiment show that the tropic effect of light in inducing a positive curvature is the same, whether it acts on the upper or on the under side of the organ.

FIG. 78. Phototropic effect of light applied at arrow on the under surface of the pulvinule.

Note effects parallel to those in fig. 77. The direction of the responsive movement is now reversed on account of application of light on the under side of the pulvinule instead of on the upper side.

The characteristic effects of unilateral action of light on the oscillating leaflet of Desmodium are:

(1) Positive phototropism exhibited by the bending of the organ towards the light.

(2) The displacement of the mean line of equilibrium towards the source of photic stimulation.

(3) The diminution of the amplitude of natural oscillation by inhibition of the movement of the organ away from the light.

Experiments were then made on the effect of the action of light on one side or the other of a stem which was executing lateral oscillations.

Effect of Unilateral Action of Light on an Oscillating Stem

The oscillations have been shown to be uniform under suitable conditions (*cf.* fig. 76); after the maximum extent of oscillatory movement to the right, the opposite movement to the left is initiated, the extent of which is equal to that to the right. This sequence of uniform oscillation is maintained under favourable conditions for a considerable length of time. In the following experiment with *Vicia Faba* the average duration of each complete oscillation was found to be about 48 minutes.

Experiment 99. *Effect of unilateral action of light on the right side of the stem.*—The specimen employed was the stem of *Vicia Faba*. In considering the record of these oscillations it is to be borne in mind that the up-curve represents a movement of the stem to the right, and the down-curve a movement to the left. Fig. 79 represents the effect of the action of

FIG. 79. Effect of photic stimulation of right side of oscillating stem beginning at arrow.

The record indicates bending of the stem towards light, displacement of the mean line of equilibrium towards the source of illumination, inhibition of movement away from light, and diminished amplitude of oscillation.

Successive dots at intervals of 3 minutes (*Vicia Faba*).

light from a 25 candle-power Pointolite on the right side of the organ. It will be noted that this gave rise to an appreciable bending of the stem towards the light as indicated in the record by the general displacement of the oscillations upwards. The magnification employed for the record was four times only; a higher magnification would have demonstrated the curvature of the stem towards the light in a more striking manner. The mean line of equilibrium was shifted upwards to the right, *i.e.* towards the source of photic stimulation. The most interesting result was, however, that the inhibitory effect of light incident on the right side prevented the completion of the natural oscillatory movement to the left. In consequence of this physiological restraint, the amplitude of the oscillation is seen to have undergone diminution.

Experiment 100. *Effect of unilateral action of light on the left side of the stem.*—The results are clearly demonstrated in fig. 80. Here, also, (1) the organ undergoes a bending towards the light; (2) the mean line of equilibrium is displaced towards the source of photic stimulation; and (3) the inhibitory action of the light incident on the left side prevents the

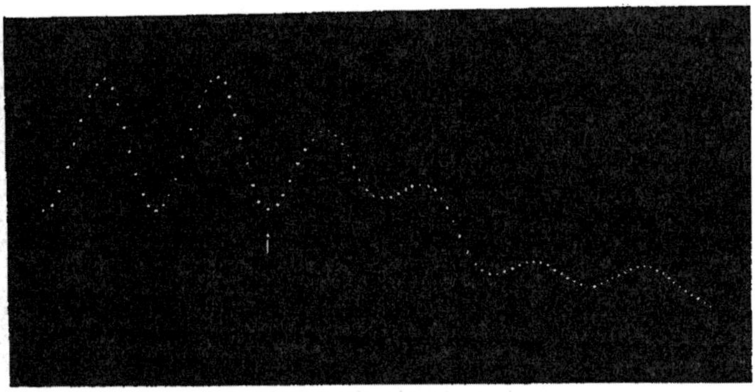

FIG. 80. Effect of photic stimulation (at arrow) of the left side of the oscillating stem.

Note effect parallel to that of fig. 79, the responsive movement being now in the opposite direction.

completion of the natural oscillatory movement to the right, the amplitude of oscillation being at the same time diminished. It may be pointed out that these effects can only be satisfactorily observed when the oscillations are uniform.

The experiments that have been described show that the action of light upon the movements of oscillating organs, whether they be leaflets or growing stems, is essentially the same.

SUMMARY

Stimulation by diffuse light causes diminution of the rate of growth of organs when in a favourable tonic condition.

Under stimulation by diffuse light, an organ in a state of

subtonicity exhibits an enhancement of the rate of growth. The tonic level of the organ is raised by the stimulation, and the acceleration of growth is then transformed into the normal retardation.

Unilateral action of light induces a positive curvature, due to the retardation of growth at the proximal and acceleration of growth at the distal side. Under longer exposure to light, the amplitude of response is increased and the period of recovery is prolonged.

Under the unilateral action of light the oscillating leaflet of Desmodium exhibits a positive phototropic curvature, bending towards the light; there also occurs a displacement of the mean line of equilibrium towards the source of photic stimulation; and, finally, the inhibition of the natural movement of the organ away from the light causes a diminution in the amplitude of response.

Parallel phototropic effects have been observed in a number of stems showing oscillating growth.

XVIII.—THE PROTEOLYTIC ENZYMES OF *CARICA PAPAYA* (I)

BY

N. C. NAG, M.A., F.I.C., AND H. N. BANERJEE, M.Sc.

INVESTIGATIONS carried out at the Bose Institute by Sir J. C. Bose on the physiological reactions of plant-extracts led us to the study of the phenomena of enzyme-action. We thus came across the accounts of the extensive pioneering work of Professor S. H. Vines [1] on the proteolytic enzymes of the plant-kingdom, who found them in pitchers, fruits, seeds, and other parts of the plant-body.

According to Professor Vines, there are two proteolytic enzymes in papain:

(1) Peptic, which he called 'peptase'—digesting proteins, such as fibrin, into peptones, but no further.

(2) Ereptic, which he called 'ereptase'—splitting peptones and albumoses to amino-acids, such as tryptophane, but not able to digest or dissolve fibrin—that is, an erepsin as defined originally by Cohnheim.

Willstätter [2] and his pupils concluded that there is but one proteolytic enzyme in papain, and that it is of tryptic nature; it peptonises fibrin, and further, when properly activated, splits peptones into amino-acids. They also found an enzyme, to which Willstätter restricts the name 'erepsin,' which differs from that of Cohnheim in that it

[1] S. H. Vines, 'The Proteases of Plants,' I to VII, *Annals of Botany*, vol. xviii, p. 289, 1904; vol. xix, p. 149, 1905; vol. xix, p. 171, 1905; vol. xx, p. 113, 1906; vol. xxii, p. 103, 1908; vol. xxiii, p. 1, 1909; vol. xxiv, p. 213, 1910.

[2] References to papers by Willstätter and his pupils are given at the end of this paper.

does not digest peptone, but acts only on the simplest peptides, such as alanyl-glycine. His view of the digestive action of papain is that (1) it digests fibrin and other higher proteins to peptones, and (2) if activated by HCN, or some natural kinase, proceeds to split peptone to amino-acids; that is, that there is only one enzyme in papain which carries out both stages of proteolysis.

In order to obtain reliable evidence of the presence of an active protease, it is essential (1) to use a suitable antiseptic so as to exclude the possibility of bacterial action, and (2) to exclude protein hydrolysis by mere chemical action. Control experiments have, therefore, to be conducted under identical conditions, side by side, with the various experimental materials, both with and without the addition of antiseptics. In the experiments, to be detailed here, these precautions were observed in every case; observations were taken at brief intervals in order to get decisive results in as short a time as possible, so as to minimise the possibility of vitiation by bacterial action.

The efficacy of the latex and juice of Papaw fruit in softening meat, dried as well as fresh, had been known for a long time both here in India and in other countries. This property is attributed to proteases (protein-splitting enzymes) present in the plant and fruit which loosen and even dissolve the fibres of the animal tissue.

In the present investigation fibrin, freshly prepared from the blood of hens, and Witte-peptone were used as the proteins for digestion by the extracts of *Carica Papaya*. The disappearance of fibrin and the development of the tryptophane reaction [1] were the digestive evidence depended upon. Suitable antiseptics were used, and simultaneous control experiments were conducted. Time taken in effecting dissolution of fibrin, and depth of the colour of the amyl-alcohol extract of tryptophane developed in the digested liquid on the addition of chlorine in slightly acid media, were the points noted.

[1] Vines, 'The Proteolytic Enzymes of Nepenthes,' I, II, III, *Annals of Botany*, vol. xi, p. 563, 1897; vol. xii, p. 545, 1898; vol. xv, p. 563, 1901. 'Tryptophane in Proteolysis,' *Annals of Botany*, vol. xvi, p. 1, 1902.

Extracts of the Leaf and of the Leaf-stalk

Extracts were made with 1 per cent. NaCl solution. The clear filtered extracts were placed in experimental bottles in incubators at temperature 39°-40° C.

(A) Extract of Leaf-stalks.

(1) 40 c.c. extract alone with a few drops of chloroform.
(2) 40 c.c. extract with Witte-peptone and chloroform.
(3) 40 c.c. extract with fibrin and chloroform.
(4) Witte-peptone solution and chloroform.
(5) Fibrin in chloroform water.

In none of the above was there any sign of enzyme activity during the 24 hours of incubation. No tryptophane reaction was observed in any of the bottles, nor was fibrin dissolved or disintegrated.

With toluol as antiseptic, the results were the same.

(B) A similar set of experiments was performed as above with extract of leaves macerated in NaCl solution. A strong tryptophane reaction was observed in the leaf-extract with Witte-peptone solution, with toluol as antiseptic, within 20 hours.

The above shows the leaf-stalk of *Carica Papaya* to be deficient in protease-content, and chloroform to be an unsuitable antiseptic.

Extracts of the Fruit

In the first series of experiments, unripe green fruits, with their rinds peeled off, were macerated to pulpy consistency, and the juice pressed through clean cloth. The clear juice was then divided into several parts and experimented with as described below :

(1) One part for autolysis; nothing added.
(2) One part with addition of antiseptic, such as toluol, chloroform, HCN, severally, for autolysis.
(3) One part with antiseptic and fibrin.
(4) One part with antiseptic and Witte-peptone.

Side by side with the above were placed for control:

(5) Witte-peptone solution in water.
(6) Witte-peptone solution in water with antiseptic.
(7) Fibrin in water.
(8) Fibrin in water with antiseptic.

Distilled water was always used, and the incubator temperature kept at 39°–40° C.

A second series of experiments was conducted with similar fruit extracts made with 5 per cent. NaCl solution. The method of experiment with these extracts was as in the previous series.

A third series of experiments was performed with papain precipitated from a 5 per cent. NaCl extract of the fruit, by treatment with sufficient alcohol (91 per cent. ethyl alcohol). The precipitate was allowed to settle, then filtered and dried. The dry white powder (in the cases where the rind was also used there was slow browning) was the material for a distilled water extract.

A fourth series of experiments was made with 2 per cent. NaCl extract of the residue left after the distilled water extraction.

The third and fourth series of experiments were on the same lines as those described by Professor Vines in his paper, 'The Proteases of Plants, VI.' ('Annals of Botany,' January 1909).

In the experiments detailed below, the antiseptic used was HCN throughout, as it was found to be more suitable than either chloroform or toluol. The quantity of HCN used was in every case half as many drops of 2 per cent. HCN as cubic centimetres of liquid taken for the experiment.

First Series.—(C) Unripe Papaw fruit was macerated to pulp, the juice pressed out and filtered through cloth.

(1) The clear juice gave no indication of tryptophane reaction to start with, but within the 24 hours of incubation and sometimes after 10 to 12 hours of autolysis there was evidence of tryptophane.

(2) The same reaction as above was obtained when 2 per cent. HCN had been added, one drop for every 2 c.c. of the juice taken.

(3) 40 c.c. of the juice with 20 drops of 2 per cent. HCN and 0·1 grm. of Witte-peptone gave strong tryptophane reaction in about 10 hours or less, the reaction becoming stronger and stronger with time.

(4) 40 c.c. of juice with 20 drops of 2 per cent. HCN and 0·1 grm. of moist fibrin, left in the incubator side by side with other bottles under experiment, gave complete disintegration of fibrin in 3 hours and complete dissolution soon after. The tryptophane reaction was the same as in autolysis (1).

Control experiments (5), (6), (7), and (8), as described on p. 179, were carried out with Witte-peptone solution alone and with HCN, as also with fibrin in water alone and with HCN. In none of the cases (5) to (8) was there any evidence of reaction.

A bottle of juice with HCl added to 0·2 per cent., gave no indication of tryptophane during 24 hours of incubation. A similar bottle with 0·5 per cent. citric acid gave tryptophane reaction, which gained in strength after 20 hours and more.

Second Series.—In this series of experiments unripe Papaw fruit, after being macerated to pulp, was extracted with an equal quantity by weight of 5 per cent. NaCl solution. A similar set of experiments as under first series (C) was performed with the appropriate controls. The results were similar to those observed in the corresponding first series experiments.

Third Series.—(D) 2500 grm. of unripe Papaw fruit pulp were treated with an equal quantity of 5 per cent. NaCl solution. The clear filtered solution was treated with 4 times its volume of 91 per cent. ethyl alcohol. The precipitate was allowed to settle overnight and filtered. The dry powder weighed 3 grm.

This was extracted with 60 c.c. of distilled water. The residue left over was subsequently used for experiment (E).

(1) 20 c.c. of the distilled water extract with 10 drops of 2 per cent. HCN did not give tryptophane reaction even after 28 hours of incubation.

(2) 20 c.c. of distilled water extract with 0·1 grm. of Witte-peptone and 10 drops of 2 per cent. HCN gave no indication of tryptophane reaction.

(3) 20 c.c. of distilled water extract with 0·1 grm. of moist fibrin and 10 drops of 2 per cent. HCN showed fibrin disintegration and complete solution in 18 hours. But there was no indication of tryptophane even after 28 hours' incubation. Biuret reaction was definite.

It is to be noted here that although the Papaw fruit juice by itself (with or without antiseptic) gave tryptophane reaction in the autolysis experiment, there was no evidence of this reaction with the papain obtained by precipitation with alcohol. In this case fibrin was digested to peptone, but the digestion did not proceed to the formation of tryptophane.

Fourth Series.—(E) The residue left after extraction with distilled water was treated with 60 c.c. of 2 per cent. NaCl solution (shaking for 30 minutes in a stoppered bottle).

(1) 20 c.c. of the saline extract with 10 drops of 2 per cent. HCN gave no tryptophane reaction even after 28 hours' incubation.

(2) 20 c.c. of the saline extract with 0·1 grm. of Witte-peptone and 10 drops of 2 per cent. HCN gave no indication of tryptophane reaction even after 28 hours' incubation.

(3) 20 c.c. of the saline extract with 0·1 grm. of fibrin and 10 drops of 2 per cent. HCN showed disintegration of the fibrin in 17 hours and complete dissolution in 22 hours. But there was no tryptophane reaction even after 28 hours' incubation.

It is to be noted that the reactions in the series (D) and (E) were in every respect similar, except that the complete dissolution of fibrin took a longer time in (E) (3). There was no difference in the nature of the digestive activity of the two extracts.

Papain prepared from Fruit Pulp with the Rind

In the series of experiments (D) and (E) the fruit pulp was obtained from the fleshy part of the fruits, leaving out the green rind. In some preliminary experiments it was noticed that the digestive action of the rind portion was stronger than that of the inner fleshy portion. Papain was therefore prepared from the entire fruits, including the green rind. The method of preparation by precipitation with alcohol was in every respect the same as that previously followed with the fleshy pulp. In the present case 3 grm. of papain were obtained from 2300 grm. of pulp. The papain in this case was not perfectly white, but showed slow browning during filtration and drying.

The 3 grm. of papain were treated with 75 c.c. of distilled water (shaking in a stoppered bottle for 30 minutes), filtered; the extract was treated as described under (F), and the residue treated with 75 c.c. of 2 per cent. NaCl solution and the saline extract treated under (G).

(F) Distilled water extract.

(1) 15 c.c. extract with 8 drops of 2 per cent. HCN did not give any indication of tryptophane reaction during 30 hours of incubation.

(2) 30 c.c. extract with 15 drops of 2 per cent. HCN and 0·1 grm. of Witte-peptone gave no indication of tryptophane during 30 hours of incubation.

(3) 30 c.c. extract with 0·1 grm. fibrin and 15 drops of 2 per cent. HCN showed complete dissolution of fibrin within 6 hours of incubation, but there was no tryptophane reaction even after 30 hours.

We must mention here that after a somewhat longer period than 30 hours (about 36 hours or more) there seemed to be some indication of tryptophane formation in experiment (2).

In some of these experiments, while testing for tryptophane in the usual way, the amyl-alcohol extract showed a strong yellow colour and not the pinkish-violet colour characteristic of the tryptophane test with chlorine water in slightly acid medium.

(G) The residue from (F) was further extracted with 75 c.c. of 2 per cent. NaCl solution.

(1) 15 c.c. of the saline extract with 8 drops of 2 per cent. HCN gave no tryptophane reaction during 25 hours of incubation.

(2) 30 c.c. of saline extract with 0·1 grm. of Witte-peptone and 15 drops of 2 per cent. HCN gave no tryptophane reaction during 25 hours of incubation.

(3) 30 c.c. of saline extract with 0·1 grm. of fibrin and 15 drops of 2 per cent. HCN gave fibrin disintegration sign within 5 hours, and the fibrin was found to have been completely digested after 17 hours. There was, however, no tryptophane reaction even after 25 hours' incubation.

It should be noted that in every case of fibrin digestion the solution gave strong biuret reaction.

(H) Observations with Papain prepared from Fruits grown at the Bose Institute

A series of observations was made with papain prepared as already described, only the fruits were taken fresh from the trees growing on the premises of the Bose Institute. The results were generally confirmatory, but with the following modifications:

Fibrin was completely disintegrated in less than 4 hours and completely dissolved in 5 hours, although the quantity of fibrin in these series of experiments had been increased to double the amount in the previous experiments.

Though the period of incubation in some of these cases

was prolonged to four days, yet there was no tryptophane reaction in any of the bottles, except a faint indication in bottle No. 2, containing distilled water extract with 0·2 grm. of Witte-peptone with 15 drops of 2 per cent. HCN in 30 c.c. of distilled water extract, on the fourth day.

NaCl solution extract of the residue, after extraction with distilled water, gave exactly similar results. Here, too, the fibrin dissolution was effected in shorter time. No tryptophane reaction was detectable even though the time of observation was prolonged to four complete days.

These last experiments show that the fruits grown on the Bose Institute premises, when used immediately after plucking, digested fibrin more rapidly than the ordinary market fruits.

Experiments in Alkaline Medium

In all the experiments so far detailed the media, in which enzyme action was studied were more or less acidic. We give, further on, the values of the different media with which we worked. We now give a few details of the results obtained when alkali was added.

(I) 30 c.c. of extract of papain (prepared by us), with addition of 15 drops of 2 per cent. HCN as in previous experiments, was made alkaline by adding Na_2CO_3 to 0·75 per cent. strength. Fibrin in such a medium dissolved completely in so short a time as 3 hours.

The solution thus obtained gave strong biuret reaction, but no tryptophane reaction.

Observations with Finkler's Papain

A sample was obtained from local chemists. Its distilled water extract was slightly coloured and acidic, coagulating on boiling, and giving biuret and xanthoproteic reactions. Generally speaking, it was much less active than papain prepared by us in the laboratory from fresh fruit. HCN was more favourable than toluol for the material, in the sense that reactions took shorter time.

(J) 4 grm. of Finkler's Papain were well shaken up with 160 c.c. of distilled water for 30 minutes. The solution filtered.

(1) 30 c.c. extract with 15 drops of 2 per cent. HCN in the incubator gave no tryptophane reaction during 24 hours of incubation.

(2) 30 c.c. extract with 0·1 grm. of moist fibrin (without HCN) did not show any evidence of fibrin dissolution.

(3) 30 c.c. extract with fibrin as above, *but with 15 drops of 2 per cent. HCN*, showed disintegration of fibrin within 16 and complete dissolution in 36 hours.

(4) 30 c.c. extract with 0·1 grm. of fibrin and 15 drops of 2 per cent. HCN as above, and with 0·5 grm. of citric acid, left the fibrin unattacked during 24 hours of incubation.

Finkler's Papain in Alkaline Medium

(5) 30 c.c. extract as in the foregoing experiments, with 15 drops of HCN and 0·1 grm. of fibrin and Na_2CO_3 to 0·75 per cent. strength, gave fibrin dissolution within 16 hours.

A similar set of experiments, when repeated with 35 c.c. of extract instead of 30 c.c., gave exactly similar results, only in the corresponding No. 5—in presence of 0·75 per cent. Na_2CO_3—fibrin dissolution was effected in 12 hours.

These results show that Finkler's Papain gave, generally speaking, similar reactions to those of the papain prepared in the laboratory, though much weaker. No tryptophane reaction was in evidence in any of these experiments with Finkler's Papain.

ACIDITY AND ALKALINITY—*p*H VALUES

The range within which the enzyme activity was best in evidence will be seen from the following *p*H values of Papaw fruit juice and papain extracts. The calomel and saturated solution of KCl, and freshly platinised hydrogen electrode method was used. The Potentiometer (Bose Institute make) was capable of reading directly to 0·5 millivolt and was sensitive to still lower variation.

(1) Fresh Papaw fruit juice from Bose Institute, *p*H value at temperature 30° C. 4·69

(2) 5 per cent. NaCl solution extract from an equal quantity of pulp, *p*H value at temperature 30° C. . . . . . 4·99

(3) Expressed juice from market fruits, *p*H value at temperature 25° C. . . . .

40 c.c. of the above juice with 20 drops of 2 per cent. HCN, *p*H value at temperature 25° C. . . . . 5·09

The foregoing made alkaline with 0·75 per cent. Na_2CO_3, *p*H value at temperature 25° C. . . . . 9·20

(4) Distilled water extract of alcohol-precipitated papain prepared in the laboratory, as used in the experiments, *p*H value at temperature 24° C. . . . . . 6·50

(5) Distilled water extract of Finkler's Papain, as used in the experiments, *p*H value at temperature 21° C. . . . . . 5·20

Distilled water extract of Finkler's Papain with alkali to 0·75 per cent., *p*H value at temperature 21° C. . . . . . 9·93

Distilled water extract of Finkler's Papain with alkali to 0·75 per cent. and HCN, *p*H value at temperature 21° C. 9·30

From the observations detailed above, it will be seen that the *p*H value of expressed juice from unripe Papaw lies about 5, being lower with fresh fruits and higher with market fruits; in any case, however, the juice is acidic. Activity had been studied over a wide range between *p*H 4 and *p*H 9 and thereabouts. Papain prepared by alcohol precipitation seems to act best in alkaline solution, nearer *p*H 9. Waksman and Davison[1] quote *p*H 5 to *p*H 7 as optimal for the protease papain. For trypsin they quote figures lying between 7·5 and 9·5; and particularly with regard to fibrin *p*H 7·5 to 8·3. They refer further to activation by HCN as being due to 'formation of a dissoluble compound between the enzyme and the cyanide.' Again, papain 'a true protease, it becomes capable of attacking peptone (peptidase) when activated by HCN.'

CHARACTER OF THE ENZYMES PRESENT IN *CARICA PAPAYA*

While the fibrin-dissolving power of the natural juice pressed out from fresh unripe Papaw fruit juice was optimal at or near its natural acidity of *p*H value 5 to 7, the extracts of the papain prepared by precipitation with alcohol, as well as those of Finkler's Papain, were most active at *p*H value much over 7 and even nearer 9. Fibrin dissolution by this prepared papain did not proceed beyond the peptone stage (except in two or three isolated long-continued experiments). What, now, is the bearing of our results upon the conclusion of Willstätter and his collaborators, that papain contains but a single protease, which is tryptic in nature, inasmuch as it digests true proteins to peptone, and, when activated by HCN or by natural kinase, splits peptones to amino-acids?

Our results do not afford any material support to the view of Willstätter and his collaborators, that papain contains but a single protease, and that this is tryptic in nature. The papain that we prepared, as also Finkler's, digested fibrin to peptone, most readily it is true in an alkaline medium

[1] S. A. Waksman and W. C. Davison, *Enzymes*, pp. 48, 224. R. Fabre and R. Frossard, 'Influence de la réaction du milieu sur la digestion papainique,' *Compt. Rend. Soc. Biol.*, 1925, vol. 92, pp. 59–60. E. H. Frankel, 'A Comparative Study of the Behaviour of Purified Proteins towards Proteolytic Enzymes,' *J. Biol. Chem.*, 1916, vol. 39, pp.

at or near pH 9; but in no case was there evidence, by the tryptophane reaction, that the peptone thus produced was split into amino-acids, even in the presence of HCN.

Although in our experiments with prepared papain we did not succeed in obtaining the tryptophane reaction, yet it must not be forgotten that the juice from fresh fruit gives the tryptophane reaction on autolysis, which is all the stronger when Witte-peptone has been added. Though, so far, we have not been able to prepare two separate enzyme-solutions from Papaw fruit, (1) one of which digests fibrin to peptone and no further, and (2) another which does not act on fibrin but effects peptolysis—*i.e.* digestion of proteoses (albumose, peptone) to amino-acids, including tryptophane—we are more in agreement with the views of Professor Vines than with those of Professor Willstätter.[1] The fact that alcohol-precipitated papain failed to digest peptone, as tested by the tryptophane reaction, although its action on fibrin was quite as active or even more so than that of the Papaw juice itself, proved the presence of a peptase—that is, a protease the action of which is limited to peptonisation of the higher proteins. Again, the fact that tryptophane was produced in the autolysis of fresh Papaw juice, and that Witte-peptone was digested as evidenced by stronger tryptophane reaction, lead us to the conclusion that there is an ereptase present in the juice, though we have not succeeded in isolating it.

We are continuing our investigations, which will, we hope, throw further light on the subject.

[1] R. Willstätter und Grassmann, 'Ueber die Activierung von Papain durch Blausäure,' *J. Physiol. Chem.*, 1924, vol. 138, pp. 184–215. 'Ueber die Proteasen der Hefe,' *Untersuchungen über Enzyme*, No. 122, pp. 1607 *et seq.* *Zeit. physiol. Chem.*, 1926, vol. 153, p. 250.

XIX.—THE PROTEOLYTIC ENZYME IN THE SEEDS OF *PHASEOLUS MUNGO* L.

BY

N. C. NAG, M.A., F.I.C., AND H. N. BANERJEE, M.Sc.

IN the course of our investigations on the composition of some of the common leguminous pulses in Bengal, and of the nature of the proteolytic enzymes present in them, we were interested to find certain peculiarities in some of them.

The seeds of *Cicer arietinum* (chick pea, *chana*, *chhola*, *gram*, *boot*) contain a higher percentage of oil than the other common pulses, proteid nitrogen being less than in some of the others.[1] An extract with water or NaCl solution gives the biuret reaction, as *also the tryptophane reaction.* If the extract be boiled and filtered, the clear filtrate still gives the above reactions. Extracts of germinated seeds give the above reactions but more strongly, tryptophane becoming more and more pronounced after incubation at 39°-40° C. Similar remarks apply to *Phaseolus radiatus*, particularly to the golden-yellow variety, *shona moog*.

Extracts of other species of pulses, as, for example, *Cajanus indicus*, *arhar*, give the tryptophane reaction only after autolysis of the extracts of germinated seeds.

The seeds of *Phaseolus Mungo* (*mashkalai*, *urd*)[2] present certain peculiarities, although their oil and proteid nitrogen content are almost the same as in those of *Phaseolus radiatus*.

[1] N. C. Nag, H. N. Banerjee, and K. Bose, 'Analysis of some Calcutta ar Pulses,' *Proc. XVIth Indian Science Congress*.

[2] Sir George Watt, in his *Commercial Products of India*, says: 'There has been some confusion regarding the nomenclature of *P. Mungo* and the species which follows—*P. radiatus*—due chiefly to Roxburgh having transposed the original Linnean names. *P. Mungo* Linn. is the present plant *urd*.'

General Reactions of Extracts of Ungerminated Seeds of *P. Mungo*

Extracts made with water or with NaCl solution gave no tryptophane reaction to start with or even after incubation for 24 hours. Whether toluol or HCN was the antiseptic did not affect the results. The extracts did not give definite biuret reaction, only a slight violet coloration being developed temporarily.

Specific Reactions of Extracts of Germinated Seeds

Working with extracts of germinated seeds in different stages of growth, the following results were obtained:

From every 10 grm. of seeds 40 c.c. of extract was made with water or with 1 to 5 per cent. NaCl solution. The mixture was macerated and filtered. Toluol and HCN were used as antiseptics in different experiments. In some of the experiments maceration was carried out after addition of a few drops of toluol. This variation in the procedure did not in any way affect the results.

So far, in none of the experiments referred to above was there any evidence of tryptophane in the extract of the germinated seeds, nor was there a distinct biuret reaction.

The following are typical observations in experiments with addition of Witte-peptone and fibrin:

(1) 30 c.c. extract with 2 per cent. NaCl solution, with 15 drops of 2 per cent. HCN, in stoppered glass bottles, gave no indication of tryptophane formation even after 24 hours in the 39°–40° C. chamber. (It should be mentioned that in a few isolated cases a faint indication of tryptophane was observed after 38 hours or more.)

(2) 30 c.c. of a water solution of 0·2 grm. of Witte-peptone with 15 drops of 2 per cent. HCN was placed side by side with (1); there was no tryptophane formation during the time of incubation, 24 hours or more.

(3) 30 c.c. of extract of germinated seeds as in (1), with the addition of 0·2 grm. of Witte-peptone and proportionate quantity of HCN, gave tryptophane reaction within 14 hours, and in some cases within 5 to 6 hours. *The action on Witte-peptone was definite.*

(4) 30 c.c. extract prepared as in the foregoing cases, but with 0·1 grm. of moist fibrin and proportionate quantity of HCN. The fibrin was not attacked during the time of incubation, 24 hours and even to 38 hours.

We would draw special attention to experiment (3), in which Witte-peptone was digested.

All of the above experiments were repeated with the addition of 0·2 grm. of citric acid: there was no variation in the results, except that the tryptophane reaction was more pronounced and was produced earlier in the Witte-peptone bottle (*cf.* (3) above).

So far the extracts were acidic in character. The effect of alkalinity of the solution was next tested.

The NaCl extract was made alkaline by the addition of 0·25 grm. of Na_2CO_3. Even then it was found that the fibrin was not attacked during 24 hours of incubation. With Witte-peptone there was weak tryptophane reaction after 24 hours' incubation when examined, but the reaction was not very definite.

Observations with acid and alkaline solutions point to the fact that extract of the germinated seeds contains an enzyme which digests Witte-peptone in acid media but has no action on fibrin.

ACTION OF EXTRACT OF GERMINATED SEEDS ON PAPAIN-PEPTONISED FIBRIN

In our paper on the nature of the proteolytic enzymes in *Carica Papaya*, we advanced the view, based on our observations, that there are two enzymes in Papaya fruit juice, one of peptase nature and the other of peptonase or ereptase nature. Alcohol-precipitated papain was found to be of peptase nature, while ereptase was in evidence in the fresh juice. Our work in this paper shows that there is ereptase, without peptase, in germinated *P. Mungo* seeds. If this conclusion is correct, it should be possible to bring about the further digestion to amino-acids, including tryptophane, of fibrin peptonised by papain, by the addition of ereptase prepared from *P. Mungo*.

The following experimental results clearly support the above anticipation:

(1) 30 c.c. of extract of alcohol-precipitated papain, as described in our foregoing paper (' The Proteolytic Enzymes of *Carica Papaya* (I) '), was made alkaline by adding 0·25 grm. of Na_2CO_3. Moist fibrin, about 0·1 grm., completely dissolved in 3 hours' time, when kept at 39°–40° C. The solution gave biuret reaction, but no tryptophane reaction, even after 24 hours.

(2) 30 c.c. of the same extract of alcohol-precipitated papain was used for the dissolution of 0·1 grm. of moist fibrin, effected in 3 hours in the incubator. This solution was then made acid by adding 0·2 grm. of citric acid and incubated for 22 hours. There was no indication of tryptophane during the period.

(3) To a similar 30 c.c. of extract as the foregoing (2), after dissolution of the fibrin in 3 hours and the addition of citric acid, there were added 30 c.c. of the extract of germinated seeds of *P. Mungo*. To start with, there was no tryptophane reaction; but after 22 hours there was definite indication of tryptophane.

In all the above experiments a proportionate quantity of HCN, one drop for every 2 c.c. of solution, was used as an antiseptic. During each of the above experiments control experiments were carried on with Witte-peptone solution, and with extract of germinated seeds, observations being taken simultaneously.

THE *p*H VALUES OF EXTRACTS OF SEEDS OF *P. MUNGO*

The following is a statement of the *p*H values of the different extracts used in the above experiments:

(1) Ungerminated finely powdered seeds. 10 grm. treated with 40 c.c. conductivity water and strained through cloth to get a colourless liquid . .	6·84 at 28° C.
(2) Germinated seeds. 10 grm. extracted with 40 c.c. conductivity water and filtered	*p*H 6·60 at 28° C.
(3) Germinated seeds with 2-in. long seedlings (3 days old). 10 grm. extracted as (2) . . .	*p*H 6·65 at 28° C.
(4) Seedlings 2½ in. long (8 days old). 10 grm. extracted as in the foregoing cases . .	*p*H 6·60 at 28° C.
(5) Seedlings 3 in. long (12 days old). Extract from 10 grm. as above	6·65 at 29° C.

It will be noted that the *p*H value of the extracts from the seedlings of different ages does not show any appreciable variation. These experiments were performed between March 7 and March 20, 1929. Repeated again at the latter end of March and in early April of 1930, the values obtained were very nearly the same.

(6) Germinated seeds were extracted with 2 per cent. NaCl solution. 40 c.c. of solution was obtained for every 10 grm.

(*a*) One part was made alkaline to the extent of 0·75 per cent. Na_2CO_3 and HCN added, 1 drop for every 2 c.c. of the extract . . .	*p*H 9·52 at 30° C.

(*b*) One part was made acid by adding 0·2 grm. of citric acid to 30 c.c. of extract and 15 drops of 2 per cent. HCN . *p*H 3·02 at 30° C.

During the course of the observations the acidity did not change or increase appreciably, except when there was putrid smell after two or more days.

Conclusion

The results of experimental investigation show that extracts of ungerminated *P. Mungo* seeds did not exhibit the activity of either a fibrin-digesting or a peptone-digesting enzyme.

Extracts of germinated *P. Mungo* seeds did not give evidence of fibrin digestion in either acid or alkaline solution, but gave definite evidence of peptone-digesting activity, particularly in acid solution, whether the material was Witte-peptone or fibrin peptonised by papain.

It appears, therefore, that the germinated seeds of *P. Mungo* contained a proteolytic enzyme analogous to the erepsin of Cohnheim, which may be termed a peptonase, since it split peptone but had no digestive action on fibrin.

These results agree in the main with those obtained by Dean [1] in his investigation of the seeds of *Phaseolus vulgaris*, but with this difference, that he found the seed of *P. vulgaris* to contain 'ereptase' even when ungerminated, whereas we found none in the ungerminated seed of *P. Mungo*.

[1] Dean, 'On Proteolytic Enzymes,' *Bot. Gazette*, 1905.

XX.—THE PROTEOLYTIC ENZYMES OF *CARICA PAPAYA* (II)

BY

N. C. NAG, M.A., F.I.C., AND H. N. BANERJEE, M.Sc.

THE results of our previous investigation of the nature of the proteolytic enzymes in *Carica Papaya* and in germinated seeds of *Phaseolus Mungo*, detailed in the two foregoing papers, led us to the belief that the so-called vegetable trypsin in the fruit of the Papaw was probably a mixture of two distinct proteolytic enzymes: the one digesting the higher proteins to peptone, which Vines provisionally termed 'peptase'; the other, limited in its action to the splitting of proteoses (albumoses, peptones) to amino-acids, which Vines termed 'ereptase.' It may be incidentally mentioned here that Grassmann[1] has recently found that the so-called trypsin of Yeast is a mixture of a protein-digesting enzyme, which he terms 'proteinase,' and of a peptone-splitting enzyme, which he terms 'polypeptidase,' which is analogous to Vines' 'ereptase.' We have, as previously stated, been guided by the pioneering work in this direction of Professor Vines. In our attempts to complete the evidence by separating the enzymes so as to obtain two distinct solutions, the one containing the peptase and the other the ereptase, we finally succeeded by the method of treatment with kaolin, a method which had already been used by Willstätter,[2] though with a different result.

[1] Grassmann und Dyckerhoff, 'Ueber die Proteinase und die Polypeptidase der Hefe,' *Zeit. physiol. Chem.*, 1928, 179.

[2] Willstätter, Grassmann, und Ambros, 'Ueber die Einheitlichkeit einiger Pflanzenproteasen,' *Zeit. physiol. Chem.*, 1926, 152.

KAOLIN TREATMENT

The Papaw fruit, unripe and fresh, was thoroughly macerated to pulp and the juice pressed out as described in our previous paper. This juice gives the biuret reaction, develops tryptophane on autolysis, dissolves fibrin and digests Witte-peptone. But if this juice be shaken up with kaolin, and the mixture either centrifuged or filtered, the supernatant liquid or the filtrate, as the case may be, no longer has any action on fibrin nor does it undergo autolysis; nevertheless, it acts on Witte-peptone, indicating the presence of ereptase. The *p*H value of the juice suffers no appreciable change on treatment with kaolin, and ranges from 5·3 to 5·6 in different samples.

The following is a typical example of the results obtained:

190 c.c. of juice expressed from finely crushed unripe fresh Papaw fruit was shaken up vigorously with 9·5 grm. of pure dry kaolin for 15 minutes in a wide-mouthed stoppered glass bottle. The mixture was then either centrifuged or filtered by pump. The supernatant liquid, or the filtrate, designated as A, gave the following indications of its nature. The kaolin sludge served for the extraction of the solution B.

REACTIONS OF SUPERNATANT LIQUID A.

(1) 40 c.c. of solution A with 20 drops of 2 per cent. HCN gave no indication of tryptophane due to autolysis during 24 hours of incubation.

(2) 40 c.c. of solution A, 20 drops of 2 per cent. HCN, and 0·2 grm. of moist fibrin. There was no sign of dissolution of fibrin nor any tryptophane reaction during 24 hours' incubation.

(3) 40 c.c. of A, 20 drops of HCN, 0·2 grm. of fibrin as in (2), with 0·75 per cent. of Na_2CO_3. Result was the same as in (2).

(4) 40 c.c. of A, 20 drops of HCN, and 0·2 grm. of Witte-peptone. Tryptophane reaction was strong within

12 hours and sometimes within even 6 hours, the reaction becoming stronger and stronger with time.

(5) 40 c.c. of distilled water, 0·2 grm. of Witte-peptone, and 20 drops of HCN gave no tryptophane reaction during the 24 hours of incubation; addition of 0·75 per cent. Na_2CO_3 made no difference.

(6) 40 c.c. of distilled water, 0·2 grm. of moist fibrin, and 20 drops of HCN. There was no tryptophane reaction even after 24 hours; addition of 0·75 per cent. Na_2CO_3 made no difference.

The above observations point to the presence of *ereptase alone in the liquid A*. Fibrin was not attacked when the liquid had been made either acid or alkaline.

B. Extract of Kaolin Sludge

The kaolin sludge was freed from adhering liquid as far as possible by filtration and use of pump, and rubbed up in a mortar with 160 c.c. of 5 per cent. NaCl solution. The mixture was centrifuged, and the perfectly clear supernatant liquid experimented with as detailed below.

The *p*H value of the liquid was 5·99 at 30° C.

(1) 40 c.c. of extract B with 20 drops of 2 per cent. HCN gave no tryptophane reaction during 24 hours of incubation.

(2) 40 c.c. of extract B, 20 drops of HCN, and 0·2 grm. of Witte-peptone gave no tryptophane reaction during the 24 hours of incubation.

(It should be mentioned here that the extract B itself gives the biuret reaction, indicating the presence of proteose. To this we shall revert later on.)

(3) 40 c.c. of extract B, 20 drops of 2 per cent. HCN, and 0·2 grm. of fibrin. The fibrin was broken up in 24 hours, but not fully dissolved. There was no evidence of tryptophane formation even after 24 hours of incubation.

(4) 40 c.c. of extract B, 20 drops of HCN, 0·2 grm. of moist fibrin as in (3) were now made alkaline with 0·75 per cent. of Na_2CO_3. There was complete dissolution of fibrin when observation was taken after 14 hours' incubation, and in other experiments after 6 hours. There was no evidence of tryptophane formation in any of the bottles. The *p*H value in this alkaline solution was 9·3, and did not change very much after solution of the fibrin.

The above experiments were repeated, using 2 per cent. NaCl solution for extraction. The results observed were similar, with sometimes longer time in the dissolution of fibrin. They all go to show that the extract from the kaolin sludge contains only a single protease which is of peptase nature, digesting fibrin to peptone and no further, and active through a wide range of *p*H values 5·5 to 9·7, more so in alkaline than in acid solution.

It may be noted here that when in some of the experiments we prolonged the time of observation to two or three days, the solutions sometimes gave a putrid odour, as well as tryptophane and acid reactions. But this never happened when the time of incubation did not exceed 30 hours.

We now give a summary of the conclusions reached from the experiments and observations detailed in our two papers on the nature of the proteolytic enzymes in *Carica Papaya*.

PAPAW FRUIT JUICE

(1) The freshly pressed-out juice from crushed unripe fruits gives no tryptophane reaction, but gives the biuret reaction, indicating the presence of proteoses.

(2) The juice, with or without antiseptics, gives on autolysis evidence of formation of tryptophane within a few hours, but the reaction is weak.

(3) The juice digests fibrin quite readily at its natural acidity, and also within a shorter time when made

alkaline; the *p*H values ranging from 5·1 to 9·7. The tryptophane reaction is then as strong or as weak as that due merely to autolysis.

(4) The juice digests Witte-peptone, as evidenced by a stronger tryptophane reaction than that due to mere autolysis (2) of the juice, or in the case of fibrin dissolution (3). This Witte-peptone digestion is quicker and stronger when the juice is made more acid. A *p*H value down even to 3 was found to act satisfactorily.

The above observations point to there being two distinct proteolytic enzymes in the juice of the Papaw fruit.

The Alcohol-precipitated Protease of Papaw Juice

(5) Extracts of the precipitate produced by alcohol digest fibrin quite easily, and particularly when alkaline. We failed to get tryptophane reaction even after 24 hours' incubation. Since the proteolytic enzyme here concerned has the power to digest the higher proteins to peptones and no further, it is evidently a peptase.

(6) Extract of the precipitate as in (5) had no action on Witte-peptone either at its ordinary acidity, *p*H 5·5 or so, or when made alkaline as in (5), the *p*H value ranging to 9·5 or so.

Treatment with alcohol thus precipitates the peptase, but the question arises: What has become of the ereptase that caused the tryptophane reaction in the juice itself or when Witte-peptone had been added to the juice? This point was subsequently settled by the kaolin method.

Separation of Ereptase and Peptase by the Kaolin Method

(7) Juice pressed out from fresh unripe Papaw fruit when shaken up with kaolin and centrifuged or filtered gave a liquid containing ereptase, as shown

by its characteristic property of digesting Witte-peptone in acid solution, *but not digesting fibrin*, even when the solution was made alkaline.

(8) NaCl solution extract of the kaolin sludge shows the characteristic property of peptase, digesting fibrin and not digesting peptones (proteoses).

In the course of these experiments on the separation of the enzymes by the kaolin method, the question arose as to where the naturally existing proteose in Papaw juice, mentioned on p. 196, had gone after treatment with kaolin, since no biuret reaction was given by the filtrate. Evidently it must have been retained in the kaolin sludge. That proteoses (albumose and peptone) are adsorbed by kaolin was put to the test by preparing Witte-peptone solutions of different strengths and shaking them up with definite quantities of kaolin. Some approximate quantitative relations have been observed. Our use of 1 grm. of kaolin for every 20 c.c. of Papaw fruit juice successfully satisfied the quantitative relation for effective adsorption. This will form the starting-point of another series of investigations which is already in progress.

The presence of two distinct proteases in Papaw fruit juice is conclusively proved by the actual separation of the two from one and the same sample of the juice by the kaolin process. We do not think that any room is left for the theory that papain is a single tryptic enzyme, one or the other of the functions of which may be inactive—a theory that has been recently propounded afresh by Grassmann[1] and criticised by Vines.[2] We therefore subscribe to the view held by Professor Vines, to whom, as also to Sir J. C. Bose, we gratefully offer our best thanks for guidance and encouragement.

[1] Grassmann, *Untersuchungen ueber die Spezifität proteolytischer Pflanzenenzyme.*

[2] Vines, *The Proteases of Plants*, 1930 (Macmillan & Co.).

XXI.—CHEMICAL EXAMINATION OF SOME INDIAN MEDICINAL PLANTS

BY

N. C. NAG, M.A., F.I.C., AND K. N. BOSE, B.Sc.

AN extensive series of investigations under the direction of Sir J. C. Bose has been in progress at the Bose Institute, on the medicinal properties of various plant-products. The medicinal efficacy of some of these had not hitherto been suspected. New methods of investigations by the Resonant Cardiograph and by the Peristaltograph [1] have revealed the physiological reactions induced by the newly discovered drugs in cardiac activity and in the peristaltic activity of the stomach. Chemical and physico-chemical examination of the plants, as well as of their extracts and ashes, must go hand in hand for a better understanding of the comparative study of their physiological effects. Over and above the difficulties inherent in the problem of the physiological efficacy of plant-products, with their seasonal variation and their modification by local conditions, there are other obstacles which have to be overcome. It often happens that there is confusion in the vernacular names used in different provinces, and even in the same province, by different collectors of medicinal plants, on whom physicians of the Indian Ayurvedic School almost entirely depend for their supply. As an example of the serious mistakes which may arise, the case of *Justicia Gendarussa* (*Bishalyakarani*) may be cited. The plant *Eupatorium Ayapana* is usually supplied as *Bishalyakarani*. In consequence of the characteristically different physiological reactions observed at

[1] *Motor Mechanism of Plants*, p. 277.

the Institute, doubt arose as to the identity of the two specimens. The present investigations have been undertaken with a view to supply such definite information as regards the chemical characteristics of the plants mentioned as would account for their different physiological action.

In the following investigation the general plan followed is to determine the amount of dry matter in the plant, the drying being effected in air-bath temperature of 100° C.–105° C. Taking the dry matter as the starting-point:

(1) The amount of ash is determined, the burning being effected at as low a temperature as possible over a Bunsen in a platinum basin.

(2) A careful examination is then made of the inorganic constituents present in the ash (as also the amount of some of these constituents present in the dry matter before burning).

(3) Microchemical tests are indicated where it was thought desirable to make them.

(4) Acidity and alkalinity of the plant-extract are determined and expressed as *p*H value. The calibrated Potentiometer is a specially constructed, calomel, saturated potassium chloride, freshly platinised hydrogen electrode apparatus, sensitive to 0·5 millivolt.

(5) Total nitrogen in connection with nitrate present in the plants is determined.

An account of the investigation of the organic principles present will be given later.

The present paper deals with the chemical characteristics of the two varieties of *Justicia Gendarussa* and of *Eupatorium Ayapana* grown at the Bose Institute and its branch at Falta on the Ganges.

JUSTICIA GENDARUSSA Burm.

yn. : *Gendarussa vulgaris* Nees.
Sans. and Hindi : *Nil-nirgandhi*, *Kala-nirgandhi*
Bengalee : *Jagat-madan*, *Jogol-modon*, *Bishalyakarani*
or *Bishkarani*.

There are two varieties of this plant mentioned by Roxburgh,[1] and we have worked with both of them. The one variety, dark purple, almost black, grows in profusion at the Bose Institute. For the other, the green variety, we are indebted to Mr. N. Mitra, Curator of the Government Botanical Gardens at Shibpur, to whom our thanks are due. The green variety was also obtained at times from the drug-plant collectors of the Calcutta market.

The habitat and medicinal use of the plant are referred to by various authors.[2]

Molly refers to its use as a febrifuge,[3] Horsfield as an emetic in Java[4]; Ainslie writes of its bruised leaves and tender shoots and decoction being used in cases of rheumatism.[5] Kirtikar and B. D. Basu[6] refer to its use as a diaphoretic, knowledge being dependent upon Indian testimony. Oil from the leaves (some say oil macerated with leaves) is used in cases of eczema, and infusion of the leaves is used internally in cephalalgia, hemiplegia, and in facial paralysis.[7] Juice of fresh leaves is dropped into the ear for earache, and into the corresponding nostril on the side of the head affected with hemicrania.[8]

[1] Roxburgh, *Flora Indica* (Thacker Spink Edition, 1874), p. 43.

[2] Kirtikar and B. D. Basu, *Indian Medicinal Plants*, No. 933, Plate 724. Nadkarni, *Indian Materia Medica*.

[3] Molly, *Hooker's Journal of Botany*, 1885, vol. vii, p. 166.

[4] Horsfield, *Asiatic Journal*, vol. vii, p. 265.

[5] Ainslie, *Mat. Ind.*, vol. ii, p. 68.

[6] *Op. cit.*

[7] Surgeon-Major Houston, *Watt's Dictionary of Commercial Products of India*.

[8] P. Kingley, *Watt's Dictionary of Commercial Products of India*.

DETERMINATION OF THE CONSTITUENTS IN PURPLE VARIETY

Moisture and volatile matter.—This was estimated by drying at the average air-bath temperature of 103° C. (100° C.–105° C.). The following are some of the results obtained :

	%	%	%
Moisture and volatile matter .	77·73	80·00	74·79
Ash	2·18	2·09	2·60

In each of the above cases the percentage of ash with respect to dry matter works out to very nearly 10 per cent. During the heavy rains the dry matter came down sometimes to 16 per cent. of the fresh leaves with 2 to 3 inches of the tender shoots. The ash-content at the same time came down to 1·8 per cent. or thereabouts. As was to be expected, the moisture value varied with the climatic changes, temperature and humidity. Precautions have to be taken, especially during the rains, to avoid dirt and silica getting attached to the leaves and shoots. However, from the results of numerous experiments, it could be concluded that dry matter formed about 20 to 21 per cent., moisture and volatile being 79 to 80 per cent.

Ash, on an average, formed about 10 per cent. or slightly higher than 10 per cent. of the dry matter.

Composition of the ash.—The following constituents were determined according to the methods indicated :

Insolubles.—SiO_2 with insolubles was determined in the usual manner by treatment with hydrochloric acid to dryness and final extraction of solubles in hydrochloric acid. The amount of insolubles shown as SiO_2 varied according to season, becoming less and less with advent of the rains and the number of prunings the plants underwent. However, during March to June the percentage of SiO_2 was on an average very near 14·5, while by September it went down to 10 per cent. or slightly above. Some actual figures are given below.

Percentage of SiO_2 in ash (March June) . . .	14·53	14·65	14·73	14·63
Percentage of SiO_2 in ash (September) . . .	10·06			

Phosphoric acid. P_2O_5 was determined by the usual method of precipitating as molybdate, and finally weighing as magnesium pyrophosphate. These determinations were sometimes checked by centrifuging the molybdate precipitate in graduated tubes against standard quantities taken side by side; the gravimetric and the volumetric methods gave concordant figures. Phosphoric acid quantity increased during the rains.

				Aug.	Sept.
Percentage of P_2O_5 .	5·32	5·62	5·84	6·27	5·84

Chlorine. The plant is very rich in chlorine, and crystals of alkali chlorides can be easily obtained on slides for microscopic examination. Cubes are most prominent with needles of nitrates. A section of a twig of the plant mounted in a drop of 91 per cent. alcohol gives the crystals within a few minutes. The method followed and found most suitable for estimation of the chlorine is the Whitehorn-Volhard method[1] (standard silver nitrate, potassium sulphocyanide, ferric alum indicator in presence of nitric acid). It must be noted here that by the method of ashing there is a certain amount of loss of chlorine, which is avoided by adding alkali to the dry matter. The amount of chlorine in ash during March to June was usually between 15 to 12 per cent., nearer the higher figure before the rains. After the heavy rains in September the figures came down to 8·5 per cent. or thereabout. The estimations carried out in March 1929 agree fairly well with some of those carried out in March and April of 1930.

				Aug.	Sept.
Percentage of Cl .	15·74	12·24	12·63	10·18	8·65

Nitrate.—The plant contains a considerable quantity of nitrate with chloride (see above). During ashing most

[1] Whitehorn-Volhard, *Cole's Practical Physiological Chemistry*, Expt. 237, pp. 245–246.

of this gets decomposed, yet there remains a quantity sufficient to give the nitrate reaction and to be estimated by nitron nitrate method.[1] For estimating the total nitrogen and nitrate in the plant before ashing a special treatment had to be adopted which is dealt with later on.

Percentage of NO_3 in ash . . . 4·07

Sulphur.—This was estimated as barium sulphate by the usual method, and SO_3 calculated therefrom.

Percentage of SO_3 . 1·57 (March to June); 2·51 (Sept.)

Iron.—Iron in ash was estimated by titration methods and gave fairly uniform results and figures between.

Percentage of Fe_2O_3 . . 1·22 to 1·30

Lime.—Calcium was estimated both by oxalate titration and by gravimetric weighing as sulphate after precipitation and filtration as oxalate (phosphoric acid having been removed). The values obtained ranged between 22·50 and 27·66 per cent. in September. More often the value was—

Percentage of CaO . 25·22 30·22 (September)

This was an extreme case.

Magnesium.—Magnesium was invariably estimated as pyrophosphate, from which MgO was calculated. Higher figures were obtained in specimens with greater profusion of leaves.

Percentage of MgO . 6·90 7·03 8·96 9·93 (Sept.)

Alkali metals.—Sodium and potassium were first weighed as sulphate, and then the amount of SO_4 deducted after precipitating this as $BaSO_4$. From the above the proportion of sodium to potassium was calculated. The following are some of the actual values:

Percentage of NaK (total) . 23·48 23·56 22·89

The proportion of sodium to potassium was about 1 to 4.

[1] M. Busch, 'Nitron Nitrate,' *Ber.* (1905), vol. 38, p. 861. A. Gutbier, *Zeit. angew. Chem.* (1905), p. 494. Treadwell and Hall, *Quantitative Analysis*, pp. 451–452.

The following is a statement of the percentages obtained in the particular samples examined:

	Purple variety		Green variety
	Before rains	During rains	During rains
SiO_2	14·53	10·06	
P_2O_5	5·32	5·84	3·29
Cl	12·74	8·56	6·45
NO_3	4·07	Very small	Very small
SO_3	1·57	2·51	
Fe_2O_3	1·25	1·20	
CaO	25·20	30·22	29·89
MgO	8·03	9·93	14·16
NaK	(total) 23·48	23·60	
Total determined .	96·19	90·92	60·82

CO_2, an undetermined amount, accounting for the difference from the total 100. There was more carbonate in the ash during the heavy rains than before the rains.

A comparison of the figures obtained for the two varieties, purple and green, shows that there was much less of silica, chlorine, and phosphoric acid, but a much larger quantity of lime and magnesia, in the green variety from the Botanical Gardens than in the purple variety.

The composition of the ash gives some idea of the composition of the plant substance. But this can only be helpful to a certain degree. It is quite probable that during the ashing the most active principle, from the physiological point of view, is lost or burnt off, so that it is necessary to examine the dry matter before ashing for organic principles that may be present. This, as already stated, will be dealt with in subsequent papers. However, we give in this paper the results of estimation of chlorine, nitrate, and total nitrogen present in the dry plant substance, as these are more allied to inorganic estimations.

Chlorine in the dry plant substance was found to be slightly higher than what one would expect to get by calculation from the ash. In an experiment in August 1929 the chlorine content in the dry matter was 0·998 per cent. The ash value was 9·95 per cent. of the dry matter. One should thus expect to get a little over 10 per cent. chlorine in ash; but the actual percentage obtained was 9·54, owing to loss during burning.

In estimating the amount of *nitrate in the dry matter*, the watery extract has to be freed from tannins and other colloidal interfering substances before the nitron method [1] can be successfully employed. The filtered extract is therefore treated with basic lead acetate, slightly warmed and filtered. The clear filtrate is made lead-free by treatment with potassium or sodium phosphate. The clear lead-free liquid is then concentrated on a water-bath, and then dealt with, after slight acidification with dilute sulphuric acid and warming, by the usual nitron method. The preliminary treatment here indicated is essential before the usual nitron method can be applied.

A series of concordant results varying between 1·22 per cent. and 1·26 per cent. of NO_3 in dry matter was recorded. One would thus expect to get 12 per cent. of NO_3 in the ash, but the actual figures are nearer 4 per cent. or even less. Evidently the NO_3 is used up during burning.

The high percentage of nitrate present lends itself easily to Molisch's test,[2] as also to the ordinary Brucine test. An approximate idea of the comparative quantities of nitrate present in different samples or species can easily be made by this test. The green variety thus shows that, comparatively speaking, it contains much less nitrate than the purple variety, which is exceptionally rich during the dry season before the rains.

It is easy to get *crystals of a mixture of chlorides and nitrates of the alkali metals from an extract of the purple*

[1] M. Busch, 'Nitron Nitrate,' *Ber.* (1905), vol. 38, p. 861. A. Gutbi..., *Zeit. angew. Chem.* (1905), p. 494. Treadwell and Hall, *Quantitative Analysis*, pp. 451–452.

[2] Molisch's test—diphenylamine in concentrated sulphuric acid—gives deep blue with minute traces of nitrate.

variety of *Justicia Gendarussa*. Analysis of a sample of these crystals gave the following result in a particular case:

Cl	24·54 per cent.
NO_3	30·23 ,,
Na	5·99 ,,
K	36·73 ,,
Total determined	97·49 per cent.

Total nitrogen in the dry plant substance was estimated by the Kjeldahl method modified by the addition of salicylic acid. It is noteworthy that the total nitrogen value in the purple variety of *Justicia Gendarussa* is exceptionally high, being 3·8 and over. A particular sample gave the value 3·88 per cent. in August 1929; while a sample examined on March 20, 1930, gave the figures as 3·86 and 3·87 per cent. The green variety is comparatively poor in nitrogen.

But the most important fact noticed in regard to this plant is not only the high nitrogen content in the plant substance, but also the alkalinity of the plant extract.

The Value of the Plant Sap

The upper portion of the plants, with about 3 inches of the tender shoot, was macerated and the sap pressed out and filtered through muslin. The *p*H value was then determined.

The following are typical results obtained:

	Purple variety
*p*H . .	8·00, March 1929
	7·86, August 1929
	7·68, September 1929

After a very heavy shower of rain the *p*H was lowest at 7·23, the green variety giving 7·10 at 30° C.

A redetermination of the *p*H values in March and the early part of April 1930 indicated a recovery towards greater alkalinity of the sap. It is to be noted that the greater alkalinity in the purple variety goes with the rather high nitrogen value.

EUPATORIUM AYAPANA VENT.

Bengalee : *Ayapan.*

A comparative study was made of this plant [1] which, as mentioned before, is commonly confused with *Bishalyakarani* (*Justicia Gendarussa*). The following results are typical of those obtained. One of the two specimens was grown at the Bose Institute, Calcutta, the other at the Falta Research Station, 30 miles down on the Ganges, with a view to study the effect of local conditions.

Bose Institute	Falta
Ash in dry matter Per cent.	Ash in dry matter Per cent.
10·99	13·28
11·04	13·28

Analysis of Ash

	Per cent.		Per cent.
SiO_2	10·23–11·04		9·26 11·82
P_2O_5	7·39		7·07
Cl	8·23		7·62
CaO	24·53–22·56		19·20
MgO	13·45		15·00–14·93
Fe_2O_3	1·09		1·15–1·22
NaK	13·84	Na	3·32
		K	10·77–13·14
SO_3	12·07		15·00
NO_3	0·10		Small quantity

NO_3 in dry matter was found to be 0·38 to 0·39 per cent. in the material of *Eupatorium Ayapana* grown at the Bose Institute.

It will be noticed there is not any very great difference, due to local conditions, between the two specimens of Ayapan. Nor was there any great difference in *p*H value.

Bose Institute	Falta
6·00, March 1928	6·00, March 1928 (when first planted)
5·83, August 1929	
5·895, September 1929	5·745, September 1929
6·08 after a very heavy shower of rain.	

[1] Kirtikar and B. D. Basu, *Indian Medicinal Plants*, No. 631, Plate 518B.

Outwardly there was no noticeable difference in the two samples. The Falta variety was, however, rather brittle. The sulphur and magnesia contents of the Falta plant were distinctly higher, while lime and NO_3 were lower. In total nitrogen the difference was more pronounced. In the plants grown at the Bose Institute the percentage of total nitrogen was 2·37, 2·79, 2·83 and near about, while the Falta samples gave values lying between 1·38 and 1·35.

In a comparative study of the various results given in this paper, the most noticeable are the high total nitrogen content of the purple variety of Justicia and its distinctly high *p*H value or alkalinity. In contrast with them are the low nitrogen content of Ayapan and its acid nature or low *p*H value. This may possibly account for the difference in the physiological action of the two plant-extracts observed by Sir J. C. Bose.

In Ayapan, Dymock mentions having found a neutral principle, for which he proposed the name Ayapanin. We have been able to prepare from the plant an organic principle which, from preliminary tests, seems to be entirely different from this Ayapanin. But the consideration of this organic constituent is reserved for a future paper.

In conclusion we have to express our grateful thanks to Sir J. C. Bose, whose physiological studies of various plant-extracts led to this investigation, and to whom we are indebted for guidance and encouragement throughout.